Municipal Knowledge Series

Stepping Up to the Climate Change Challenge

Perspectives on local government leadership, policy and practice in Canada

Edited by:
Susan M. Gardner and David Noble

brought to you by the publishers of

CANADA'S MUNICIPAL MAGAZINE SINCE 1891

2008

©Municipal World Inc., 2008

All rights reserved. No part of this publication may be reproduced, stored in a retrieval system, or transmitted, in any form or by any means, photocopying, electronic, mechanical recording, or otherwise, without the prior written permission of the copyright holder.

Library and Archives Canada Cataloguing in Publication

Stepping up to the climate change challenge: perspectives on local government leadership, policy and practice in Canada / edited by Susan M. Gardner, David Noble.

(Municipal knowledge series)
ISBN-10: 0-919779-86-7

ISBN-13: 978-0-919779-86-0

1. Climate changes—Government policy—Canada. I. Gardner, Susan M. II. Noble, David III. Series.

JS1710.S84 2008   363.738'740971   C2007-907454-5

Published in Canada by
Municipal World Inc.
Box 399, Station Main
St. Thomas, Ontario N5P 3V3
(Union, Ontario N0L 2L0)
2008

mwadmin@municipalworld.com
www.municipalworld.com

ITEM 0095
Municipal World — Reg. T.M. in Canada, Municipal World Inc.

Printed on

TABLE OF CONTENTS

FOREWORD iv

Chapter 1
IT'S A WICKED PROBLEM 1

Chapter 2
OVERVIEW
A New Look at the Big Picture 7

Chapter 3
LOCAL LEVEL POLITICS OF CLIMATE CHANGE 15

Chapter 4
TOP DOWN, BOTTOM UP, ACROSS, INSIDE OUT
A Climate of Change in Local Government 21

Chapter 5
CHANGING THE POLITICS OF BUSINESS-AS-USUAL . . 31

Chapter 6
ADAPTATION THROUGH RISK MANAGEMENT 37

Chapter 7
EDUCATION FOR SUSTAINABLE DEVELOPMENT 43

Chapter 8
INTEGRATED COMMUNITY SUSTAINABILITY PLANNING
A Process for Addressing Climate Change
at the Local Level 49

Chapter 9
COMMUNITIES OF TOMORROW 57

Chapter 10
WHO HAS THE ANSWERS?
A Small Community Perspective 63

Chapter 11
COLLABORATING TO ADDRESS CLIMATE
CHANGE IN EDMONTON 71

Chapter 12
A UNIQUE MULTIFACETED APPROACH
Response of the Montreal Public Health Authorities
to Climate Change 77

Chapter 13
GOING GREEN 85

Chapter 14
PLANNING FOR CHANGE IN RICHMOND HILL 91

Chapter 15
USING CLIMATE SCENARIOS AT THE
MUNICIPAL SCALE 97

Chapter 16
GREEN MUNICIPAL FUND
Facilitating Local Learning and Action 105

EPILOGUE 111

ENDNOTES 115

ABOUT THE EDITORS AND CONTRIBUTORS 117

RESOURCE SECTION 121

LIST OF FIGURES

Figure 4-1: Governing Meta-Structure 27

Figure 6-1: Risk Management Process 38

Figure 8-1: Sustainable Community Quality of Life 54

Figure 12-1: Montreal Urban Heat Island Map 79

Figure 15-1: Direct and Indirect Impacts 98

Figure 15-2: Adaptive Capacity of a System 101

Foreword

Anne Golden

There is no sensible reason to delay action on climate change. As the fourth report of the Intergovernmental Panel on Climate Change makes clear, the effects of climate change are already with us, and will worsen unless we take immediate and concerted action to avoid the most serious outcomes. To delay will not only increase the costs of reversing negative trends, but will also lengthen the time required to do so.

Recognizing climate change as both an environmental and economic issue is essential, particularly as we think about devising strategies for our cities. With four in five Canadians living in urban areas that are centres of industrial development and major drivers of our economy, cities are a key focal point in addressing climate change. Cities need to understand how they contribute to climate change and what their vulnerabilities are, so that they can take appropriate steps. Doing so will not only yield environmental benefits, but will also protect their economic interests. This book makes an important contribution to the dialogue, sharing information from and actions taken by Canadian communities. The progression from past and current municipal experiences, through shared challenges, to available programs, tools and resources is both logical and thought-provoking.

The most widely-understood component of climate change is mitigation of greenhouse gas emissions. It is not surprising, therefore, that more and more cities are seeking to reduce their carbon footprint. Efforts such as the C40 Large Cities Climate Summit, the Cities for Climate Protection, and Zerofootprint Toronto are focussing attention on how we plan and develop our urban areas, and the implications for our hydrocarbon-dependent energy system.

Actions to mitigate greenhouse gas emissions are typically better understood than measures to adapt to a changing climate. This lack of awareness of the need for adaptation is a concern, as cities are particularly vulnerable to natural hazards, and the IPCC suggests that disaster-generating events are likely to increase in both number and intensity. The aging state of urban infrastructure raises the likelihood that we will not be sufficiently equipped to deal with extreme weather-related events. We should also be concerned about the human health effects.

Resilient and sustainable communities come about through forward planning and a thorough understanding of challenges and possibilities. Climate change is, in many ways, a game changer that demands new ways of doing things. Our prosperity as a nation depends critically on the success of our major cities, and our cities need more resources to deal with traditional issues of land use, urban sprawl, transportation congestion, and industrial development. Each of these issues impacts mitigation and adaptation to climate change. How municipal leaders respond to this challenge will shape the lives of most Canadians in the years to come. This volume records some of those responses and describes the context within which further actions will be taken.

Chapter One

It's a Wicked Problem

David Noble and Trevor Dixon Bennett

The verdict is in!

Once upon a time, not so long ago, there was a lively debate about climate change. Was it really happening? Were humans really causing it? If the wheels of global climate change were in motion, could we really influence the extent of change and its effects?

The years 2006 and 2007 may go down in history as the years when the climate change issue reached the tipping point. The debate over these questions is over. The answers to all three are resoundingly *yes*. Climate change is real, it is here, and it is serious.

In 2007, thousands of climate scientists from countries all over the world, appointed by their governments and collaborating through the United Nation's Intergovernmental Panel on Climate Change (IPCC), concluded that "warming of the climate system is unequivocal," and that the warming of the planet is "very likely"[1] due to human activities.

Around the same time, the American Association for the Advancement of Science (publishers of the journal *Science* and one of the world's premiere science institutions), released its statement on climate change: "The growing torrent of information presents a clear message: we are already experiencing global climate change. It is time to muster the political will for concerted action. Stronger leadership at all levels is needed. The time is now. We must rise to the challenge. We owe this to future generations." This is a remarkable statement for what is traditionally a conservative institution espoused to objectivity.

Sir Nicholas Stern, former World Bank Chief Economist and one of the world's pre-eminent economists, described climate change as "the greatest market failure the world has ever seen." Stern warned that inaction could lead to a drop of 5 to 20 percent in global GDP, the displacement of 200 million people, and "swaths of the Earth's surface becoming uninhabitable." Oil giant Chevron Texaco's Chairman declared, "Climate change is, without question, the single greatest environmental challenge we face." Former UN Secretary-General Kofi Annan said, "It's not just an environment issue, it's an all-encompassing threat."

Right here at home, Sheila Watt-Cloutier, former Chair of the Inuit Circumpolar Conference, in referring to the already severe impacts of climate change on the Inuit in the Canadian Arctic, now speaks of the tragedy with the language of human rights. Lloyd Axworthy refers to the situation in Canada's north as a cultural genocide.

Climate change also hits home in cities and city life. That's what cities like Stratford, Ontario will tell you. Stratford now faces a $250-million class action lawsuit for failing to anticipate heavier rains. Many other cities that have experienced extreme storm events, prolonged heat waves, and damaged infrastructure, will agree. London (UK) Mayor Ken Livingston notes that in cities, "we see the problems, we see the damage that carbon emissions are doing, like the threat of flooding and the violent weather and terrible levels of heat."

Individually, Canadians are on board. In early 2007, public concern for the environment (read: climate change) trumped concerns for the economy, security, and health for the first time in 30 years of tracking. By mid-summer, 91 percent of Canadians indicated they should do their part in the fight against climate change, even if doing so would cost them more.

We can only conclude that climate change is real, it is here, and it is now.

It Isn't Going Away Any Time Soon

The wheels of climate change are already in motion. It is a phenomenon that is already stressing ecosystems and human populations in Canada and around the world.

And, no matter how successfully we reduce GHG emissions to the atmosphere, the wheels are going to continue to roll. As a result of the GHGs that are already in the atmosphere, our climate will continue to warm for at least the next two decades – even if we were never again to emit a single GHG. Given our continued reliance on fossil fuels, it is likely to get much warmer yet.

The climate impacts that we've experienced to date – the storms, the floods, the heat waves, the droughts, and all of their associated effects – are likely only a sampler of what is yet to come. Indeed, the worst is still to come. Even if the political attention on climate change wanes, the ecological crisis that's unfolding – and need to deal with it – won't.

The debates about whether or not climate change is happening are all but dead. Now, the debates focus on how much, how fast, what to do about it, and how to get it done.

The Problem Is ... It's A Wicked Problem

In a perfect world, the atmosphere would harmlessly absorb all the waste we dump its way. There would be no climate change. If only the world were perfect.

In a second-best world, we would simply and easily do what is required to reduce GHGs, and adapt to the changes to which we are already committed. Climate change would be a pest, but we would deal with it, quickly and relatively painlessly. If only the world were even second-best!

In the real world, it's not so simple. Because there are so many complexities and dilemmas inherent in the climate change issue, solving it isn't so easy. In the early 1970s, planning theorists coined the idea of a "wicked problem" to describe complex problems with various characteristics that made them rather difficult to deal with. While the wicked problem concept long preceded our present focus on climate change, in many ways, it captures perfectly the challenges of addressing the issue. Climate change is a wicked problem!

Ask 10 people, "What is the climate change problem?" Likely, you will get 10 different responses. Ask again six months later, and you will probably get a few more. Their responses will depend on many factors – their worldviews, their professional roles and responsibilities, their ages and incomes, whether they live in floodplains, whether they've seen the devastation of the mountain pine beetle on BC's precious forests, whether their kids have asthma. The problem – the *real* issue – depends on who you ask.

Whatever the problem is, it may well be only a symptom of another problem. Is the problem that climate change is causing more severe flooding in the floodplain, which is causing more and more property damage? Or, is the problem that local governments haven't enacted or sufficiently enforced floodplain restrictions? Maybe the people that live in the floodplain can't afford to move out because the local economy is depressed. Or, maybe all the well-to-do are building riverside mansions for their scenic views. Don't they know of the risks? What are they thinking? Maybe its an awareness problem, or an education problem, or a culture problem.

The problem is ill-defined. Necessarily, then, so too is the solution. As Dr. Jeff Conklin (author of the book *Dialogue Mapping: Building Shared Un-*

derstanding of Wicked Problems) puts it, "since there is no definitive 'the problem,' there is no definitive 'the solution'."

There are no objective criteria that identify when *the* or even *a* solution is found. There is nothing to say, for example, that if we reduce greenhouse gas emissions by 80 percent by 2050, we will "solve" the climate change problem. Is a 10-metre setback in the floodplain the right answer? Wouldn't a 12-metre setback be better? We can only ever say that one option is "better" or "worse" than another, or "good enough" or "satisfying." No matter what we decide on as a solution for today, we may decide otherwise tomorrow – as we learn more about the problem, as our values change, as our interventions change the problem we are trying to address, or perhaps as our interventions unexpectedly cause other problems. And, who gets to decide? Democracy reigns, after all – aren't we all equally entitled to judge a solution?

Whatever we decide, we have to get it right. Our aim as planners and decision makers is to improve some characteristics of the world, of people's lives. If we fail to get the solutions right, we aren't doing our jobs. We have no right to be wrong.

Nor do we have the luxury of experimenting. Every solution we try has consequences. Solutions cost money, they can irreversibly affect people's lives, and they may spawn new problems. These effects can't easily be undone. Climate change already kills some 150,000 people globally each year. We can't bring them back. The same goes for the species that go extinct. At its worst, climate change will be truly catastrophic. We have one shot at preventing it. Life as we know it is at stake. With just one shot at preventing catastrophe, we must get it right.

Cue the Cultural Shift

To say that climate change is a big issue is to put it mildly. Its implications are massive. As with most things in life, the response and solution should be proportionate to the problem. Given the scope and characteristics of global climate change, the solution can be nothing short of a massive cultural transformation.

This will require major changes across all facets of our existence: from changes in individual lifestyles, to changes in global capital flows, and everything in between. As Montreal Mayor Gerald Tremblay says, "Cities are part of the problem, so they must be part of the solution." Cities must change and become more sustainable. Greening or transforming our cities can have a huge impact on aggregate emissions levels. These kinds of changes are absolutely crucial for climate protection.

If a cultural transformation sounds like a daunting task, that's because it is. At the same time, though, it is entirely possible. Culture embodies and reflects our patterns of behaviour and activity, our habits, our beliefs, our technologies. Our culture can be influenced; it is dynamic. Indeed, we make little changes to it with every decision we make.

We are already seeing these changes in our communities, albeit on a limited scale. Communities across Canada and around the world are pioneering a cultural transformation. Sustainability wasn't a consideration in local planning 25 years ago and, before this decade, green power procurement policies barely existed. Now, new forms of partnership – like the one between the City of Regina and Home Depot to promote home energy efficiency, and profiled in this book – are challenging our traditional ways of thinking about public service delivery. These and many other developments represent new ways of conducting municipal business, all aimed at being more sustainable. This is cultural transformation in real time.

Many European cities are much further along in their push toward sustainability. While their progress can be a sobering reminder of how far we have to go, they provide an inspiring and comforting reminder that we can get there. In 2006, Canada's Commissioner on Environment and Sustainable Development, Joanne Gélinas reported on the federal government's climate change programs and activities. The Commissioner called for "a massive scale up of effort." Her words are equally fitting to local governments: A massive scale up of effort is required.

News to Norms

By way of example, Guelph city council recently approved a community energy plan. The plan sets ambitious per capita goals for electricity, water use, and emissions reductions, with the ultimate goal of transforming Guelph into a more sustainable city. The initiative was championed by a community consortium comprised of many key partners, including the city, the local power utility, school boards, the University of Guelph, and others. The city is quickly moving into major demonstration projects that will build momentum for large-scale implementation. The community is on board. It is all very exciting.

Federal Liberal party leader Stéphane Dion visited Guelph to discuss the plan and to celebrate its success. He remarked ... what is happening in Guelph is exemplary, but hopefully it would soon be standard practice.

Like the Guelph community energy plan, the experiences and initiatives presented in this book are exemplary. They offer examples of best practices and lessons learned; they benchmark the leading edge of municipal sustainability; they are worth sharing

As exemplary as these stories now are, the hope is that they will quickly become the norm. Climate change *is* a wicked problem, but it is not one without solutions – and those solutions are within the grasp of local governments in Canada.

Chapter Two

Overview

A New Look at the Big Picture

David Noble and Aiden Abram

If you look at the science that describes what is happening on earth today and aren't pessimistic, you don't have the correct data. If you meet the people in this unnamed movement and aren't optimistic, you haven't got a heart.

~ *Paul Hawken*

Climate Change 101

For such a massive and complex problem, the basics and recent history of climate change are remarkably simple.

1. Over the course of 200 years of industrial development and (mostly) unsustainable living, greenhouse gas (GHG) concentrations have increased in the atmosphere. The level of carbon dioxide (CO_2) in the atmosphere has grown to 380 parts per million (ppm), a nearly 40 percent increase from pre-industrial levels. Concentrations of two other GHGs, methane and nitrous oxides, have also increased considerably. These increases are largely from burning fossil fuels (CO_2 is a by-product of fossil fuel use), deforestation (trees store CO_2 that is then released into the atmosphere), and various agriculture practices (including the billions of cows that, over time, have "emitted" huge amounts of methane into the atmosphere). Collectively, the GHGs now in the atmosphere far exceed concentrations seen anytime in the past 650,000 years.

2. GHGs in the atmosphere work much like a greenhouse. Heat from the Earth that would otherwise escape into space is trapped[2] by GHGs in the atmosphere. This is a good thing – if not for the greenhouse effect, our planet would be much

colder than it is, and life on Earth would not be possible. The more GHGs in the atmosphere, though, the more heat trapped, the warmer Earth gets. That's exactly what we've seen over the last century – the mean temperature on Earth has increased by 0.74°C; in Canada, the mean temperature has risen by more than 1°C. Since 1995, and probably not coincidentally, we have experienced 11 of the 12 warmest years on record.

3. Greenhouse gases last a long time in the atmosphere (CO_2, for example, has an atmospheric lifetime of 200-450 years). Because of this, we are locked into an estimated 0.1°C of warming in each of the next two decades, from the GHGs already emitted. Since we continue to emit GHGs, we can expect warming of at least another 0.2°C per decade. It's going to continue to get warmer – possibly much warmer – before it gets better.

4. As the Earth gets warmer, average and extreme temperatures tend to increase. This has all sorts of implications for people and cities. For example, warmer temperatures will tend to result in shorter, milder winters and longer, hotter summers, and more frequent, more severe, or longer-lasting heat waves. We've already experienced these changes. Glaciers, like the ones that are the primary source of drinking water across parts of western Canada, are melting faster than they are replenishing. Like a bank account that is being drawn down faster than it is being topped up, it is only a matter of time before the glacier runs out. This example (and it is only one of many) raises concerns that the Prairies will be "drier than a dustbowl."[3] More really hot days (intensified in cities by the urban heat island effect) are, at best, discomforting and kill only a few people. At worst – like Europe's 2003 heat wave that killed at least 35,000 people – they are downright toxic to quality (and quantity) of life.

5. The warmer atmosphere causes the hydrologic cycle (the movement of water, in its various forms, around the Earth) to speed up. That can cause more frequent and more explosive storm events, like hurricanes and extreme rain events. These, in turn, can result in more floods, landslides, and many other hazards that are bad news for people and property.

6. Changes in temperature and precipitation regimes cause broader environmental changes that have all sorts of implications, some

good, but most bad. Mosquitoes and ticks, for example, can move into new habitats where they would not have survived previously. Along with them, they bring the West Nile virus and Lyme disease. Other critters, like the mountain pine beetle, are able to survive the warmer-than-they-used-to-be western winters. Because of that, beetle populations are exploding and are ravaging British Columbia forests. They have already crossed the Rocky Mountains into Alberta, and there is little preventing them from spreading across Canada. These are only a few of many examples. Taken together, the environmental implications are mostly bad news for Canada and Canadians.

Beware of 2°C!

Two degrees Celsius is accepted by many climate experts and governments, including all European Union member states, as a critical threshold. If we warm by more than 2°C above pre-industrial temperatures, irreversible and potentially catastrophic climate changes become far more likely.

Global temperature increases of more than 2°C have massive implications worldwide. For example, at 2°C, 15 to 40 percent of species face extinction, and 40 to 60 million additional people become exposed to malaria in Africa. Jump to 4°C, and agriculture yields decline by up to 35 percent, and some regions (parts of Australia, for instance) suffer complete crop failure. At 5°C, higher sea levels threaten low-lying major world cities such as New York, London, and Tokyo.[4]

The climate has already warmed by 0.74°C, and will continue to warm by at least 0.1°C per decade (more likely by 0.2°C per decade) due to the GHGs already in the atmosphere. We are well on our way to 2°C.

Climate scientists estimate that, to limit the likelihood of catastrophe, we need to limit atmospheric GHG concentrations to between 450 ppm and 550 ppm[5] in the medium term, and then reduce concentrations over the longer term. We are still largely dependent on fossil fuels for powering our factories and fuelling our cars, and we still have lots of cows, despite their bad habits. Greenhouse gas emissions continue. The concentration of GHGs in the atmosphere is increasing by about 2.5 ppm every year. The path we are on could have us at 550 ppm as soon as 2035. (If you are under the age of 50 or anticipate being especially long-lived, this is probably within your lifetime.) By century's end, the average global temperature is projected to be 2 to 4.5°C warmer, with a best estimate of 3°C.

The worst is not inevitable, though. Business-as-usual is a choice; a more sustainable development path is also a choice. By (very significantly) changing the ways we produce and use energy, we can mitigate catastrophic climate change and prevent the worst it would bring. By adapting to the changes we don't or can't prevent, we can reduce their adverse effects. The choices won't make themselves, though; we have to make them. And, we have to follow through.

All Hands on Deck

There are lots of prescriptions for how to achieve the major emission reductions that are needed to avoid the worst. These include conserving energy, generating energy from renewable energy sources, capturing and storing carbon underground, taxing carbon emissions, trading carbon credits ... the list goes on and on. Successful long-term emission reductions, on the scale that is required, will therefore require everyone to do everything.

China and India will need to come on board. So will the United States, and Europe, too. And so will Canada, emitter of just over 2 percent of all GHGs globally, and one of the highest emitters on a per capita basis.

Municipalities will need to figure prominently in the mix. More than 75 percent of GHG emissions worldwide occur within municipal boundaries. They originate from buildings operated by municipalities, on the roads that municipalities plan and build, and in the homes and businesses that are at the heart of community life. Still more emissions originate outside of municipal boundaries, generated in the course of supplying municipal with energy, products, and services. Through their operations and purchasing decisions, through their policy and planning decisions that impact on community life, and through their citizen engagement, municipalities have a huge impact on GHG emissions. Municipalities have a crucial role to play, if for no other reason than because they can.

If more reason is needed, then look to self-interest. With their "vast assets in facilities, parks, roads, bridges, waterfronts, dykes, water and sewage networks, and a stake in the prosperity and safety of citizens that depend on this infrastructure, local governments are on a collision course with climate change."[6] It's worth saying again: climate change hits home in cities and community life!

Fortunately, municipalities aren't alone. Many organizations are committed to supporting, in various ways, municipal efforts to engage successfully on climate change.

The Federation of Canadian Municipalities, for example, operates the Partners for Climate Protection program (in partnership with ICLEI Canada – International Council for Local Environmental Initiatives), the $550-million Green Municipal Fund, and the recently launched Green

Municipal Fund capacity building program. Together, these initiatives offer direction and support to municipalities engaged on climate change, and sustainability more generally.

Since the early 2000s, Natural Resources Canada has supported more than 25 research projects focussed on cities and communities. These have helped build knowledge and tools that will hopefully prove useful Canada-wide.

Across the country, the provinces are also delivering programs and services to municipalities. In Alberta, for example, two provincial government departments partnered to deliver the ME First! program, a $100-million interest-free loan program to help municipalities increase energy efficiency and develop renewable energy.

The Institute for Catastrophic Loss Reduction is a not-for-profit research institute that helps communities to become more resilient. It is presently developing an enterprise risk management program for natural disaster risk reduction in Canadian municipalities.

As an example from the many across the private sector, 2degreesC has worked with *Municipal World* to help deliver climate change articles since early 2006. Readers have indicated the articles have raised awareness, offered interesting and useful perspectives, and practical examples from which they can learn. This book, which evolved out of that partnership, is a further effort to support municipal engagement on climate change.

Headfirst Into the "Messy" Middle

The many pervasive implications of climate change for Canadian municipalities (and for society more generally), naturally draws interest and activity from across society. All orders of government, the private sector, civil society, and concerned (and even unconcerned) citizens all have interests that may be affected by climate change. Given the scope of the problem, it can't be understated: we all have a role to play.

But, the rules that shape our respective roles and responsibilities are anything but clear. Local governments have no explicit mandate to tackle climate change, save for the more organic mandates emerging from the commendable efforts of municipal and community leaders. There are lots of "tips" for individuals, businesses, and other organizations, but we are far from having a clear understanding of how to be a part of the solution.

Ideally, all sides are at the table and working toward common goals. The public sector, for their part, would have a "whole-of-government" approach, in which the resources, expertise, and authority of all orders of government are deployed in a coordinated fashion – after all, as Clive

Doucet rhetorically reminds us in Chapter 5, don't all orders not represent the same people?[7]

All of this makes for a complex and "messy" institutional arrangement. With many players at the table, coming from different backgrounds, with different ways of understanding,[8] and with different values and priorities, we shouldn't expect otherwise. This messiness isn't unusual – it is a recurring and systemic feature of any complex environmental problem. It is simply the nature of the beast. The ability to make sense of and manage this messiness is critical to dealing effectively with the climate change problem.

Hear, Hear!

Like Paul Hawken in the opening quote to this chapter, you might feel pessimistic – and with good reason. The problem is huge and complex – wicked, that is! And, as acknowledged in Chapter 1, the massive cultural transformation required seems a daunting task. Despite the challenge, though, there is reason for hope.

Municipal leaders from across Canada and around the world are rising to the occasion. In December 2005, more than 300 mayors and municipal leaders from 37 countries committed to reduce GHG emissions by 30 percent by 2020 and 80 percent by 2050. More than 150 Canadian municipalities and 600 in the US have signed on to nationwide initiatives to reduce GHGs. Forty of the largest cities worldwide have partnered in the C40 initiative, which has already attracted $5 billion from the Clinton Foundation to support municipal energy retrofits in 15 of those cities.

Provinces, national governments around the world, businesses, professional associations from different sectors, Canadian and international youth – all are making commitments to help address the climate crisis. Taken together, their commitments are a significant first step toward a new way of doing things, toward a massive cultural transformation.

From Promises to Pragmatism

But, commitments are just words, and without actions, aren't going to solve the climate crisis. How do we move climate change commitments to real actions and measurable results? How do we make the changes we need to make, in a way that satisfies our various accountabilities, all the while sustaining the normal provision of municipal services?

Unfortunately, there are no failsafe formulas. But, there is a growing body of experience on how local governments and others are getting it done. None have had all the data they would have liked; none had answers to all of their questions (they likely didn't even ask all of the questions to which they would have liked answers); and none likely had all the resources they needed. Theirs are the stories of motivated and dedicated municipal and community leaders who have charged forward on the path to sustainability. Their successes speak for themselves.

And, although the climate crisis is new, many of the underlying challenges are not. Local governments have dealt with other environmental and social problems in the past. They have lots of experience managing the "messiness" and getting it done. They know the importance of leadership, of successful change management, of innovation and risk-taking, of getting started. Much of this experience is relevant to the climate change challenge.

Reflect on your own experiences with complex issues, and the experiences of your peers. Then, go ahead and get started. Don't wait until you know it all – you never will.

By engaging your partners in a dialogue, you can develop a shared understanding of the problem, and a shared commitment to addressing it. Maintain a learning orientation, not just as an individual, or as an individual organization, but as a network of partners working together. By learning together, you will make sense of the problem and inch closer and closer toward agreeable and practicable solutions.

At the heart of the solution is leadership. As Kevin Lynch, Clerk of the Privy Council observed, "Leadership isn't about working longer hours, or harder, or taking on more responsibility. No, it is about engaging your partners, setting the agenda, taking risks and being a role model."[9]

The climate crisis is putting government leadership (at all levels) to the test. Local governments are stepping up, whether through international initiatives like the C40 initiative, or through local-level initiatives like Guelph's community energy plan, or any of the experiences depicted in the chapters ahead. These are a great start. They are shaping the agenda, and providing models to emulate. Let's applaud their successes, but get on with the work ahead.

Chapter Three

Local Level Politics of Climate Change

Devin Causley

The last year has seen incredible coverage of climate change in the media, in political discourse and in the scientific research community. All of these dialogues echo one consistent message: decisive leadership on climate change is urgently needed.

Whether seated in the federal, provincial or municipal order of government, elected officials determine the direction and speed that our society moves. They make the decisions on programs, projects, budgets, and policy. Yet, elected officials are seldom trained in climate change science, let alone complex issues that span regional and international boundaries.

So, how do elected officials make decisions on climate change? What factors affect the decision-making process? These questions are the focus of this chapter, which incorporates data from a recent research survey (the survey).[10]

Canadian Municipal Context

At the moment, federal engagement on the climate change issue is stalled and made worse by frequent changes in political leadership in recent years. (Over a seven-year period, Canada has had four climate change plans.) If the federal role remains vacant, then provincial and municipal governments will need to fill in the vacuum and work together to create the leadership required by all orders of government.

Municipal governments are already demonstrating leadership at the council and staff levels. Their mandates can provide essential support to other orders of government in the implementation of policies and projects, whether through transit services, waste management (landfills, recycling), or land use planning. Their close connection to the public education system also supports local outreach and community education initiatives. Early on, municipal engagement occurred as a result of the economic opportunities presented by mitigation (i.e. energy efficiency, and renewable energy generation). More recently, the focus has shifted to the prevention of legal and economic damages as a result of climate change impacts.

Climate change will also have impacts on local economic development. Survey responses reveal that many municipalities feel their local economy is vulnerable to the impacts of climate change. The nature of the expected impact is less certain, however. When asked how they believe climate change will impact them, the majority believe there will be a combination of positive and negative impacts, while a significant number didn't know what impacts to expect. In response to anticipated impacts, half of all municipalities surveyed had developed a climate change action plan in an effort to mitigate future impacts and adapt to the expected impacts. But, decision makers require more information on local economic impacts in order to effectively integrate climate change strategies into economic development planning.

Respondents to the survey stated needs that both complement and go beyond those noted by the Federation of Canadian Municipalities in its report on "Enlisting Municipal Governments in a National Approach to Clean Air and Climate Change." A nearly unanimous response to the question of what municipal councils require to address climate change locally finds increased funding from the provincial and federal governments at the top of the list. Close behind are calls for changes to provincial or federal legislation, improved provincial and national policy, along with training and education for the municipality. Information on resources and issues, improved voter support, and the acquisition of in-house staff expertise are also found to be areas of opportunity to support municipal efforts. While responses indicate a need for further resources to be directed towards municipalities, there is also a clear message: further action on their behalf is being inhibited by policy, legislation, and funding mechanisms controlled by the other orders of government.

Governance

The ability of municipal governments to effectively engage on the climate change issue is dependant on institutional capacity and the presence of local champions. While champions typically have the entrepreneurial skills to advance the issue, they may lack formal knowledge. Instead, local champions are cultivated by other individuals who possess the knowledge and passion, but lack the authority or entrepreneurial skills to take broader action. Champions, especially at council, must be able to frame the climate change issue in a local context. This can be done by linking climate protection to the core goals of the municipality, or to issues of a more immediate local nature (i.e. economic development).[11] When this is done, the science of climate change becomes less important.

Where the environment is a prominent issue, municipalities have had greater success in addressing climate change by treating it as a policy response to local issues like air quality or quality of life, rather than an individual problem. Air quality, energy management, and climate change planning are seen as connected issues; thus, a practice of integrating these into a single planning process is emerging. This practice broadens the message to achieve greater support by decision makers, while not only avoiding potential conflicts in the implementation of strategies, but also making more efficient use of scarce staff and financial resources.

Political champions are critical to get community-wide programs beyond institutional thresholds; however, they also tend to suffer from greater turnover compared to staff. Successful engagement of a municipality can be dependent on the ability of champions to institutionalize the activity, so that it remains after the initial champion is gone (for example, by creating staff positions or semi-independent agencies).

Urban and rural municipalities experience different challenges when engaging on climate change. Urban regions benefit from a greater pool of staff resources, knowledge, and skills, and oftentimes closer, better access to external resources (eg. universities). At the same time, however, the decision-making chain often becomes longer and more cumbersome, and there are more demands competing for council support.

Rural municipalities, on the other hand, suffer from a lack of scale, and often an uncertainty about how they contribute to the problem. How does a community of 500 people, one municipal building, two vehicles, and a few light bulbs contribute to climate change? On their own, the potential contribution of such communities to mitigation is minimal compared to the urban giants. But, when working together with other small municipalities nearby, or on a region-wide scale, the contribution becomes more significant. A regional approach can reduce the burden and political risk for any one council, while reaching economies of scale for implementing practical projects.

The Challenge

Municipal governments face two distinct challenges in climate change planning: establishment of a process, and survival through electoral cycles. Where the political will exists, staff positions can be created, often in collaboration with a community organization, to support a champion that can initiate the collaborative process of climate change and energy planning.

Time and accountability present unique challenges to council, however. While mitigation measures such as landfill biogas capture or building retrofits can be implemented within a single council term, the results may not be noticed before the subsequent election. This can present obstacles for staff recommendations on future or ongoing projects to a new council. Regular updates to council and an education strategy for newly elected members can help ease the concerns of council.

Issues of accountability can also be a challenge. A council is elected to support the interests of its local community. Although it may have a conscious interest to support a regional or international movement to address climate change, this requires local evidence of a need to act – either a significant contribution to the problem (emissions) or consequences from the impacts. This results in voter support.

Like other environmental programs, when budgets are constrained, climate change initiatives may be considered a luxury. Higher up-front investment costs can be a budgetary barrier. Municipal budget planners may be short-term thinkers and may resist strategies that see the municipality investing in energy efficiency and recouping costs from the measures to fund future projects (performance contracting). Budgets are often approved on a step-by-step basis, rather than as a package, lowering risk but increasing overall cost. Cost effectiveness, economic benefit, and the quantification of results are necessary to justify expenditures and report on impacts. Thus, an initial investment in creating an institutional business case for action may be required, with support across departments and from council.[12]

Influencing Municipal Decision Making

A diverse range of stakeholders have a vested interest in municipal decisions on climate change. These include: community groups and the electorate, local business and industry, the not-for-profit sector, and the media. An informed electorate and community groups can be solid supporters for council. Citizens must be prepared to help educate their council on the impact municipalities have on the environment, while supporting council during critical decisions.

Local business and industry has a role, and is often more supportive than expected. Economic activity in the construction, farming, finance, trade, tourism, and insurance sectors is sensitive to climate.[13] Climate change poses a threat to industry that, through mitigation, can actually improve competitive advantage.

The not-for-profit sector brings support to council from within the community and beyond through regional or national associations. These groups can help network and build capacity within the organization. Local organizations can provide human resource support, research data, and act as facilitators for a climate change action team. At the same time, national organizations can support advocacy on policy, funding, and legislative issues, and can provide a broader network (eg. Partners for Climate Protection).

Finally, the media acts as a public filter to the scientific world on climate change. The media has a major role in propelling or slowing the message on climate change generated by scientists. Media reports highlight the latest disaster or warning, but seldom focus on the resistance of specific groups to taking action. These groups are often easy to identify, and their motives easily documented. Media outlets may also be controlled or influenced by groups with a vested interest in slowing substantive changes.[14]

Making Decisions

Information can be delivered to municipal decision makers from internal and external sources. The most credible and influential are those internal sources, often within the municipality (i.e. staff reports), or regularly involved with the municipality (and thus seen as an institutionalized stakeholder – i.e. municipal associations, other orders of government). External groups will find it difficult to gain a favourable decision from council if their position is not supported by municipal staff or institutional groups.

A traditional view of decision making by municipal government has been that citizens vote in elections, but after that the chosen representatives have the final decision on individual issues. In such an environment, decisions are fast, but often risk greater scrutiny by the public and media.

Collaborative decision making is becoming more common. Through task forces, roundtables or civic forums, interest groups, experts, and the general public can be engaged on climate change, and thereby contribute to local capacity. Council also benefits from the creation of a base of support to defend expenditure on climate change initiatives. To succeed, these mechanisms require political and administrative support, along with at least a minimal financial commitment.[15] Financial commitment establishes greater credibility along with an expectation of results, while avoiding such processes being used as a means for council to avoid decision making – through the deferral of sensitive issues to an endless committee process.

When asked about what influenced their decisions on climate change, municipal decision makers indicate that highly influential sources include municipal staff and federal/provincial government programs or policies. Modest influencers include the local electorate, community groups, and scientific reports, while the least influential bodies include the media, major employers, and the local chamber of commerce or economic development agency. Economic development agencies hold low influence on climate change decisions due to the limited connections that have been made between climate and economic development. There is, however, a need to involve these groups more in the process of mitigation and adaptation.

Moving Forward

Do municipal governments have a role in addressing climate change? Absolutely! The survey indicates that three-quarters of respondents believe their municipality has a significant, direct role to play in combatting climate change. An even greater percentage believes their municipality also has a role to educate their broader community. To date, municipalities have answered the call through energy efficiency, renewable energy, and innovative planning.

Moving forward with substantive reductions in GHGs and adaptation in our communities will require Canada's local government leaders to be visionaries, risk takers, and collaborators.

FCM's report on "Enlisting Municipal Governments in a National Approach to Clean Air and Climate Change" notes that an integrated approach between governments, even within Canada, will be challenging. Legislation set by other orders of government, limited access to capital (dependency on property taxes), and a lack of monitoring and assessment protocols for environmental quality at the local level are only a few of these challenges. All orders of government can support municipal leadership by enhancing clean energy incentives (efficiency and renewable energy sources), adapting infrastructure to climate change, and engaging in public education.[16]

Additional Reading and Resources

H. Bulkeley and M. Betsill (2003), *Cities and Climate Change: Urban Sustainability and Global Environmental Governance*, Routledge, New York.

Harcourt et al. (2007), *City Making in Paradise*, Douglas & McIntyre.

M. Roseland (2005), "Chapter 15: Lessons and Challenges," *Toward Sustainable Communities*, New Society Publishers, Gabriola Island.

L. Lundqvist and A. Biel (2007), *From Kyoto to the Town Hall*, Earthscan, London.

Chapter Four

Top Down, Bottom Up, Across, Inside Out

A Climate of Change in Local Government

Alex Boston

Municipalities are acutely vulnerable to climate change because of their great concentrations of people, infrastructure, and economic activity. Three quarters of GHGs are emitted within municipal boundaries. (The remainder are from industries like oil and gas, mining, agriculture and forestry that take place predominantly in territories without municipal organization.) Local government decisions inadvertently influence half of Canada's emissions.

Local governments are awakening to their vulnerability and their power to mitigate climate change impacts. A chorus of hundreds of municipalities has declared a commitment to deep emission reductions.[17] Yet, only a few dotted across the country have really begun to translate these words into action.

In Canada, a dozen local governments have stabilized emissions in their internal operations. Several have made double-digit reductions with plans to cut emissions as much as 50 percent during the Kyoto commitment period. All are finding it difficult to bend the emission trajectory in the community at large. Some, nevertheless, are making dramatic reductions here, too.

Strong business cases underpin successful projects. Drivers include energy cost savings and renewable energy income. Co-benefits include better air quality and enhanced community livability. Most of these emission reduction practices are well known and well promoted.

If the benefits are so great, and the practices so established, why are so few local governments reducing emissions? The primary reason is that conventional local government institutions, such as organizational structures and planning processes, constrain action. Local governments are unintentionally designed for building GHG-intensive communities.

Leading municipalities, however, are transforming these institutions. They are establishing new governing systems that extend planning horizons, and advance priorities, top-down, bottom-up, across the organization, and into the community. Global climate change is positioned to resonate with local priorities.

This chapter examines the institutional changes that underpin some of the leading local government climate programs.

The Real Challenge: Institutional, Not Technological

Today, we have the technologies to cost effectively cut emissions in half in a generation. Our challenge is not technological; it is institutional.

However, in endeavouring to advance cutting edge technologies, the prevailing approach is to promote best practices. It is consistent with orthodox economics, which suggests that people act rationally to maximize advantage on the basis of full and relevant information.

This orthodoxy, however, is misleading. Studies of best practice programs reveal that policy implementation rarely flows from the provision of technological knowledge.[18] In sum:

concern + best practice knowledge = action [low efficacy]

While technologically instructive, best practices often fail to account for the unique local social, economic, and institutional context. Studies of large firms and local governments responding to high profile best practice programs with strong business cases have underscored the need to "focus on organizational concern and capacity."[19]

A more accurate assessment of the change process can be summarized as follows:[20]

concern + best practice knowledge + *institutional capacity* = action [high efficacy]

In Canadian municipalities, staff and elected officials that understand institutional change have been much more effective in establishing strong local climate change programs. They have established structures and processes that out live their tenures. Those that have been unable to hardwire climate change management into organizational design have seen their efforts wither.

Institutional Challenge for Local Government

In business and in government, new policy agendas challenge well-established institutions. The climate change issue strikes to the core of these challenges.

Upfront investments in energy efficient infrastructure, renewables and many other climate protection measures are often higher than status quo infrastructure investments. Local governments, in particular, confront acute capital shortfalls, due to the revenue constraints established, in part, by other orders of government. Short electoral cycles and annual budget pressures further contract the financial horizon. This short-term orientation has significant implications for GHGs and for long-term capital and operating costs, which are generally lower with low carbon investments.

Local governments are organized around line departments, each with unique policy norms. While necessary to deliver services for a functioning municipality, inter-departmental interaction and community engagement are traditionally limited. Although conventions are changing, the climate change issue demands exceptional cross-departmental collaboration, and considerable community interaction, stretching the capacity for change. A community energy system, for example, requires significant coordination between planning, engineering, and finance, as well as collaboration with community partners, including real estate developers, utilities, energy companies, the financial services sector, and others.

Municipalities, like any institution in the private or public sector, possess tremendous inertia. Staff and citizens perform their daily activities according to well-developed patterns. Intricate systems – from training to job descriptions, and social expectations to fiscal incentives – reinforce this behaviour. If a significant policy change does not address these more systemic issues, change is often minimized or thwarted.

Positioning Climate and Strengthening the Bottom Line

While threatened polar bears and melting ice caps capture headlines, engagement from council, the bureaucracy, and the community grows when climate change dovetails with local priorities. In studying US municipal success on global climate change, Michele Betsill referred to this as a "think local, act local" approach.[21]

The strongest programs become part of a powerful narrative that resonates with community values. For example, when Sudbury was pulling together its climate plan, employment was waning in the forestry and mining sectors. Staff calculated that the city as a whole spent $400 million annually on energy; the overwhelming majority of those dollars left town. With active stakeholder engagement, its climate plan became a community economic development strategy, with a target to reduce dependence on outside energy sources by 50 percent. Investing in efficiency and renewables has created local jobs and kept more money circulating in Sudbury. This local economic narrative captured a strategy that powerfully resonated with local priorities.

In Toronto, climate change has dovetailed with air quality concerns. In Vancouver, the focus has been livability. In Calgary, it was about fiscal performance and being number one.

Staff and council stewarding successful programs generally see the climate change issue as an unprecedented opportunity to address multiple local priorities, integrating social, economic, and environmental imperatives. Good measures can reduce energy bills, ease congestion, establish fiscally sustainable infrastructure decisions, generate revenue from renewable energy, and, as mentioned, strengthen local economic development, air quality, and community livability. Plus, they reduce GHGs.

Engagement: Staff, Stakeholders, Citizens

No small group inside city hall has the intellectual, political, and financial capital to design and implement a strong climate plan. Richard Binder, who manages Calgary's climate program explains, "Twenty-five years ago, city staff not only didn't listen to citizens, they didn't talk to their colleagues in the department next door. Today, the more we talk and the more we listen, the better the decisions are that we make."

The rationale for engagement is threefold: stronger plans, enriched understanding about the challenges and opportunities, and increased sense of ownership. The engagement imperative is the same inside government and out in the community.

Some of the more important principles include thinking long-term, acting short-term; engaging end users and opponents; using efficient and effective processes; and respecting and responding to input.

The best ideas in Fredericton's corporate plan have come from a committee of 25 volunteers from across the organization. Assistant city manager, Chris MacPherson explained his team was well aware that they were saving money and reducing GHGs when they began calculating long-term energy costs in the budget for constructing new buildings. During a brainstorming session, they conceived the idea of incorporating the first five years of fuel costs along with the capital costs of new vehicles. With vehicle right-sizing, this practice transformed their vehicle fleet.

Similarly, staff suggested that the lesson of saving money and energy from high efficiency LED traffic signals should be applied more broadly. Now, service and emergency vehicles have flashing LED roof lights, enabling them to be turned off when stationary. Fredericton's rich staff engagement process allowed the city to stumble upon these solutions before they were in vogue.

Vancouver has earned a strong reputation for public engagement. Thousands of citizens were involved in defining a livable city and creating a vision to densify and improve quality of life in the downtown core. The process generated pedestrian-first and complete neighbourhood policies, and laid out a sustainability vision exceeding staff's imagination.

The result: despite a population increase of 50,000 over the last decade, vehicle numbers are stable, and vehicle kilometres traveled are down 30 percent. Transits trips have increased 50 percent, and walking and biking trips have doubled. Integrated transportation and land use planning have positioned Vancouver as one of the only Canadian jurisdictions within reach of achieving its Kyoto target. Engagement made it possible.

Innovative Financing and Strategic Management

Local climate managers identify financing as their greatest barrier. Ralph Torrie, who has worked with municipalities on climate change mitigation since the early 90s, notes that "the barrier is only partially explained by capital shortfalls."[22] Institutional norms often compromise strategic investment.

For example, split incentives emerge when the office managing construction is not the same as the one operating the building, and the ultimate decisions are made under year-end budget crunching pressures in the finance department where staff are encouraged to hold a "first cost" orientation. The result is that buildings cost a bit less to construct, but significantly more to operate. The City of Hamilton calculated that construction costs account for only eight percent of a civic building's cost over its 30 to 40 year life cycle. Operating costs – such as maintenance, repairs, and energy – comprise 92 percent of overall costs.

Ironically, the fiscal constraints and burgeoning infrastructure debts that municipal governments confront are created in part by an urban design approach that fails to take into account the full range of costs over time. The result is that local governments generally generate less in development fees and property tax for low density, single-use urban design over the long term than is spent on services like emergency and waste removal, and infrastructure costs such as roads, watermains, and sewers.

Successful climate programs hurdle these institutional barriers with management tools like lifecycle analysis, full cost accounting, and design charettes that bring key stakeholders together to share their interests. These tools help extend the planning and financial horizons beyond three- to five-year simple paybacks, also using practices such as performance contracting, revolving funds, and leveraged partnerships with other public, private, and non-profit partners.

Toronto's Better Buildings Partnership (BBP) illustrates this financing approach. Projects normally have several financing partners: a bank (whose high interest loans are paid back first), an energy service company (paid back from savings), and then the BBP. The BBP finances up to 30 percent of a project with a low interest loan that is paid back last. This small boost guarantees deep retrofits, and reduces a 10-year return on investment to five, making projects highly appealing.

The BBP has saved local businesses $19 million in energy bills, and will retrofit 40 percent of Toronto's commercial floor space within the Kyoto period. Annual GHG reductions will be the equivalent of taking 550,000 vehicles off the road permanently.

Calgary exemplifies how a corporate inventory can become a dynamic performance management tool. Most local governments carry out static GHG inventories, taking an emissions snapshot at a particular point in time. Initially, they are useful to help identify priorities and establish a baseline. However, an inventory that helps manage emissions over the long term is ideally dynamic.

Calgary constantly collects billing data for electricity, heat, and transportation fuels, along with solid waste and sewage data in a central inventory. This system is able to generate a real time account of GHG emissions, as well as common air pollutants, such as sulphur oxides, nitrogen oxides and particulate matter, along with energy and waste expenditures. This performance management system enables targets to be constantly and strategically established, monitored, evaluated, and adjusted.

Calgary has scheduled 700 civic building retrofits, reducing energy bills by an estimated $7 million annually; its wastewater treatment plant operates on its own methane; and 37 new wind turbines will soon power city operations. Calgary will cut its internal corporate emissions 50 percent within the Kyoto period, and at the same time strengthen its fiscal performance. Calgary embodies the axiom: if you don't measure it, you can't manage it.

Reinventing City Hall

In explaining Vancouver's progress in advancing sustainability, former planning director Larry Beasely said, "We're reinventing city hall." Many of these institutional changes across leading municipalities are reflected in new organizational designs. Although each is homegrown, they can be described by a governing meta-structure as shown in Figure 4-1.

Chapter 4

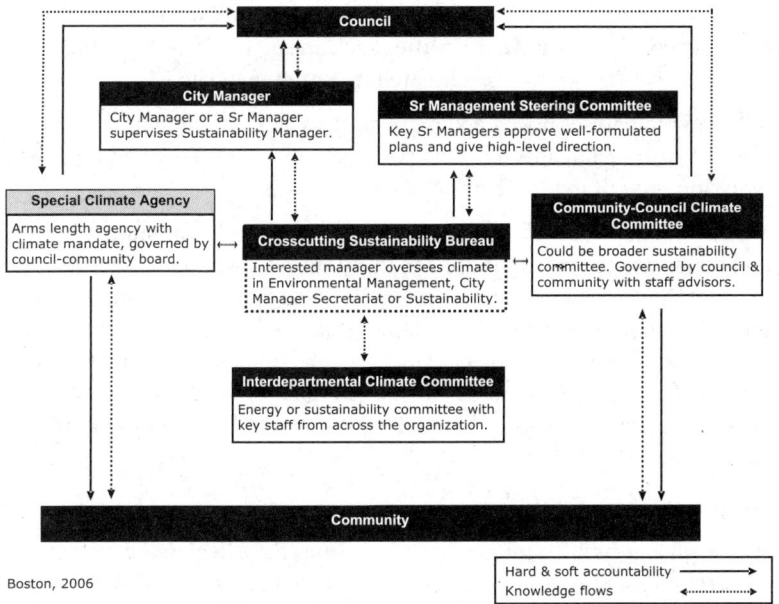

Figure 4-1
Governing Meta-Structure for Climate Protection

Boston, 2006

Hard & soft accountability →
Knowledge flows ←→

Almost all strong programs have a central coordinating bureau even if they are staffed by one part-time person. In Fredericton, it is in the city manager's office. In Regina, it's in corporate affairs. Vancouver has a separate sustainability group. While the form varies, the function is the same: a strong mandate and the authority to work horizontally across line departments.

With scarce resources and an interest in institutionalizing change, these cross-cutting bureaus are predisposed to coordination and facilitation over implementation. They favour integrating priorities into existing line departments, rather than initiating new ones. They emphasize adding value over adding volume.

Most leading municipalities engage council and senior management in high-level bodies that provide strong mandates, integrate priorities across the organization, and drive them vertically top-down. Many have interdepartmental committees to strengthen planning and implementation, driving priorities bottom-up. Strongly mandated bodies with council and community representation have been created in some circumstances to guide and support community climate protection. They may have a broader perspective, such as a focus on "sustainability," rather than just "climate protection."

In some communities, arms-length special climate agencies, with boards composed of community and council members, are established. The Toronto Atmospheric Fund exemplifies such an agency. It has significant investment capital and a clear climate protection mandate.

Most municipalities have been more successful in strengthening their internal government capacity. This is reflected in stabilizing and reversing emission growth in their corporate operations.

However, about 95 percent of emissions are in the community at large, and this is where emissions are generally surging. Municipalities that are building governance structures and processes that harness the power of the community are better at containing and reducing these community emissions. These institutions tap into the intellectual, social and financial capital of property developers, utilities, neighbourhood associations, universities, colleges, NGOs, financial institutions and other key community stakeholders.

Conclusion

Although a growing chorus of municipalities have declared commitment to deep emission reductions, only a few have been successful in significantly reducing their emissions. Staff and elected officials in these municipalities underscore that their greatest challenge is not technological; it is institutional. They have, in turn, developed new structures and processes to overcome the inherent institutional constraints to building low-carbon communities.

By thinking locally and acting locally, these municipalities have repositioned global climate change so it resonates with council, staff, and the community. Climate protection is approached as a way to advance the municipality's triple bottom line.

Staff, citizens, and stakeholders are meaningfully engaged. This helps strengthen plans, foster ownership, and enrich implementation. Innovative financial and program management approaches have made climate protection a strategic investment priority.

A governing meta-structure has emerged. A central bureau is established to coordinate across line departments. Council and senior management are engaged in bodies to drive priorities top-down. Interdepartmental committees drive priorities bottom-up.

While not as common, council-community bodies tap into community capacity. Special climate agencies, arms length from government, are highly effective in some places to finance and mobilize climate protection activity in the community at large.

Notably, those few governments moving beyond their own operations to make significant progress on their community-wide plans have been successful in two regards. First, they have built a much more robust governance capacity that taps into the power of their communities. Second, they are applying the same analytical and financial tools and management practices that are used internally, to their community-wide programs.

No local government has figured out everything. But, the communities discussed in this chapter have important lessons for municipalities beginning to tackle climate change. All leading local governments can learn something from one another.

Chapter Five

Changing the Politics of Business-As-Usual

Clive Doucet

When I was first elected to Ottawa council, I thought the city needed better transit and environmental policies. I still believe this to be so; but, I gradually came to realize that the problem wasn't about "devising" better transit and environmental policies. People have been writing cogent, incisive books about how to create cities that are less environmentally damaging for 60 years. Jane Jacob's famous *Death and Life of Great American Cities*, for example, written almost 50 years ago, comes immediately to mind. William H. Whyte's *Lost Landscape,* written 60 years ago, is another. The list is a long one. You could fill libraries with such books. We have lots of knowledge; it isn't knowledge that is the problem.

What we need is politicians and political systems that can respond in a useful way to the knowledge, and a new vision of what constitutes the "good life" and a "great city." Until we, as a society, decide that the preferred urban environment is not more malls, more parking lots, and more suburbs in green fields, nothing is going to change. We will keep doing exactly what we're doing now – building out the urban footprint, many times faster than our population growth.

Second Fiddle to the National Drama

Like every Canadian city, Ottawa was founded in the colonial era and is dotted with buildings and customs that recall those days. But, while other Canadian cities quickly matured into their own places, with their own voices in the local and national debate, Ottawa remained anchored in its colonial past.

Colonel By, of Bytown fame (the pre-confederation name for Ottawa), couldn't imagine any authority other than Westminster, and for good reasons. It was where his funding and authority came from. The men who actually built the dams, sluiceways and canals, and fell like flies to malaria, overwork, and under nourishment – but went on to found villages and towns along the canal route – had about as much political importance as a team of horses.

This early sense of colonialism and its matching imperium were deeply rooted in the little village of Bytown. Not surprisingly, this sentiment was easily transferred to the new Canadian federal government when it arrived, simply replacing Westminster. This filial attachment has never really changed for many of Ottawa's citizens.

It shouldn't surprise anyone. The overwhelming presence of the national government bumps and grinds into every aspect of local life. With its immense land holdings, its unbridled ability to purchase and dispose of property, to employ people, to build local bridges and roads, and to set the local cultural agenda simply by imposing its own, the federal imprimateur dominates the character of the city. On the evening news, "Ottawa" is mentioned and the Parliament buildings appear for the viewer. This is the side of Ottawa life that residents do their level best to politely ignore. But, it is more difficult for local council members to avoid when faced with watching the city's future decided elsewhere – because, after all, that's council's job.

A Fairer Share of the Taxes

Ottawa's situation is extreme, but it is certainly not unique. All Canadian municipalities are prisoners of a Constitution written in the days of the horse and buggy – when 80 percent of the nation was rural, not urban. Our communities have changed, but our governance systems haven't.

As a result, all municipalities, including the largest ones, suffer from a tremendous political imbalance between the local and the federal/provincial governance levels. More than 92 cents of each Canadian tax dollar goes to the national and provincial orders of government. Municipalities have only the remaining eight cents to provide 60 percent of the public services that Canadians use.

This is not new. Cities have been crying out against the fiscal imbalance for years. What is new is that climate change is advancing so quickly it will soon melt down local government's ability to secure a reasonable quality of life for urban residents.

It was fascinating to witness the very quick federal response to the federal-provincial fiscal imbalance once it was deemed to be a priority of the Premier of Quebec. Two billion dollars were quickly sent to the province. Yet, the federal-provincial imbalance is trivial compared to the imbalance between municipalities and the federal level of government, and the gap is growing with each passing year. By contrast, the federal/provincial gap is shrinking – and all by itself, because both levels have tax bases that grow with the economy.

How can we ever achieve sustainable growth when the federal and provincial governments' treasuries increase with population growth, but municipalities, who provide most services, see their costs increase and net revenues shrink? It is a recipe for business-as-usual.

Vision Matters

Every North American municipality is now struggling with the consequences of 60 years of building communities that respond to people's visions of the suburban utopia – the bungalow, the picket fence, the "Leave it to Beaver" kind of street that is a safe haven from all the ills of the world. This is no longer a recipe for a better community or a better life. The great irony is that this comforting vision has become a driving force behind the planet's great malaise – climate change.

Ottawa suffers more than most Canadian communities from the suburban ideal, because it has had the National Capital Commission (NCC), a federal agency of immense power, throwing its financial, planning, and regulatory weight behind it. The NCC has been working diligently for more than 60 years to erase any messy urban reality from Ottawa neighbourhoods. The famous Jacques Greber's plan was all about tidying up Ottawa and creating "a capital Canadians could be proud of." This phrase is used over and over again in NCC films and literature about "improving" Ottawa.

Creating a capital that Canadians can be proud of has proven to be a long and complicated task. A thumbnail sketch would include: getting rid of all of the smoky and dirty industries; getting rid of the old clanky streetcars and rail lines; creating suburban office parks for civil servants; physically eradicating working class neighbourhoods by expropriation, and then bulldozing the houses into the ground; and decorating the entire city with fine parkways in every direction, to replace the trains and clanky streetcars. These parkways, with their lovely vistas, are part of a national capital that "Canadians can be proud of."

Like any good colonist, Ottawa city councils have been all too supportive of this suburban vision. It's a long and comfortable tango. For example, when the federal government offered to buy city buses to get the old streetcars off Wellington Street, the city happily took them up on the offer. Thus began the process of dismantling one of the largest, oldest, greenest streetcar systems in North America. The Ottawa streetcars were designed and built in Ottawa, were fuelled on green power from the Chaudiere Falls, and served all the principal streets of the city.

The result: in 1960, more people per capita rode Ottawa's streetcars than today ride the billion dollar busway. The city's population has increased by 40 percent in the last three decades, but its road system has grown by 75 percent. And, in spite of all the ink the climate change crisis is getting, nothing has changed. In fact, it's getting worse. In 2007, for example, Ottawa grew its road system by 200 kilometres – an asphalt growth record.

Again, Ottawa's case is extreme, but not unique. Nearly all Canadian cities have been seduced and reduced by the suburban vision. You can't find a city in Canada from St. John's to Victoria that isn't ringed by 60 years of building car-dependent suburbs.

Modern cities exist because of the fossil fuel wealth of our planet, which has permitted us to ship foods thousands of kilometres and drive casually in every direction for the slightest of reasons. For councils like Ottawa's (and others right across the country), it was easy to divest our cities of all those robust, complex, pedestrian-based communities, that previously formed the backbone of every Canadian city. The populations in our cities were small, the air was clean, and there were no environmental bills to pay.

But, times have changed. Our vision of the "good life" and the "good city" have gotten us into the mess we're in. We need a vision to get out.

Local Governments to the Rescue

Our choices are very clear. As a society, we can decide to move fast and furiously to change the way we live so that we don't continue to deplete the planet's natural capital. Or, we can wait until climate chaos forces change. One way or the other, it will happen. We can be sure of that.

Smaller light bulbs and smaller cars are important. But, the greatest levers available to us are all public. Individuals cannot build urban rail lines or intercity rail. Individuals cannot create local advantages for local farmers, require new buildings to have green roof construction, and mandate that old neighbourhoods become carbon-neutral urban environments. Only governments can do these things.

If the federal government cared about climate change and how it affects people:

- ▶ It would move far more aggressively on policies and projects that would make a real difference to our carbon imprint now.

- ▶ It would respond to the reality that more than 75 percent of GHG emissions originate within municipal boundaries and, as a consequence, it would have in place a national green housing strategy and an urban transit strategy. (We currently have neither.)

- ▶ Rail would be a priority across the country. There would be fast passenger trains between all our major cities, and a comprehensive freight service from small town Canada to the major cities. None of this exists. There is less rail service in 2007 than there was 50 years ago.

- ▶ It would restructure the nation's taxes by transferring GST and income taxes to Canadian municipalities, enabling them to manage the growth and responsibilities being forced upon them.

It is clear that the federal government will not lead on these vital issues. It is too ponderous and too divorced from the lives of communities and ordinary people even to understand the gravity of the situation that we are all facing.

I once asked a prominent journalist in a public forum why he paid so little attention to city politics, and his response was bewilderment. "You have no national party. You're just local. If you want national attention, you need to form a national party." I remember his response many years later because it expressed so clearly the difficulty of breaking out of the current political paradigm.

All Canadian cities have to shake their colonial status and start behaving as if they were important. (This was one of the last public statements of Jane Jacobs before she passed away.) Councillors must stop thinking that a seat on council is a step to becoming a provincial or federal member of Parliament. We have to start behaving as if climate change matters – because it does. And we have to start behaving as if the most important place to be is where we are – because it is. Three-quarters of GHGs are produced within municipal boundaries, where local government provides more than 60 percent of the services Canadians use.

The emerging climate crisis is all around us. Asthma is now the number one reason we admit children to hospitals and it's related to the orange/sulphur sheen to the sky, which you can see on almost any day when it isn't raining. Scientists tell us the Bow River Glacier on which Calgary depends will be gone in 10 years. It will join the glacier of the eternal snows of Mount Kilimanjaro, which is already gone. World sea levels are rising. This means some of our traditional urban habitats are no more, or are going. New Orleans is an example. These problems demand our attention.

These are the cards that we are playing with, and we're playing with our eyes shut.

Let Us Lead With Courage and Imagination

Mayors everywhere must start behaving like the mayors of London, Stockholm, Paris, Toronto and Montreal. These big city mayors have figured out there is a crisis happening around them. Canadian municipal councils need to find the same courage to reduce traffic, reduce road expenditures, and move their communities away from single occupancy vehicles.

We need to act on what the knowledge brokers are telling us: business- and politics-as-usual is no longer sufficient. The Sierra Club says that three- to five-storey buildings are the most environmentally friendly. These buildings don't require elevators to be habitable, can be successfully green roofed, are large enough to be dense, but small enough to create interior communities. It will take courage to say to developers, "Folks, you're going to make less money, but that's tough. Our collective future is more important than your short-term bottom line."

We must have the courage to say to the federal and provincial government, "If you abandon your financial commitments for public transit or refuse to develop any national urban strategies to mitigate climate change, we will find a way without you."

The profoundness of this change in political attitude is best captured in a letter signed by the mayors of the planet's great coastal cities, and then sent to President George W. Bush during the first term of his presidency. The letter asked him to sign the Kyoto Protocol. I found a copy of it at the World Social Forum in Porto Alegre. It brought tears to my eyes when I saw the mayor of New Orleans's name affixed to it. It was written years before Hurricane Katrina descended on his city. It just seemed so sad. Why should one level of government have to beg another for the policies to prevent catastrophe? Do we not represent the same people?

Above all, we need to recognize that our political system is broken, and we need to fix it. In the era of climate change, the old partisan, political games no longer make any sense. There are answers to the climate crisis, but they won't be found in hydrogen fuel cells and other Hail Mary passes. They will be found by transforming our politics-as-usual approach into something that will respond to the needs of our time.

Chapter Six

Adaptation Through Risk Management

Dan Sandink

If you want a resilient community, you need to be proactive. Governments traditionally address natural hazards as they occur, responding to needs as opposed to proactively reducing impact. There is a common misperception that natural hazards and the environment are static and linear, with the same characteristics from year to year, over a long time frame.

Management of natural hazards has typically involved analyzing specific hazards, implementing a hazard-specific solution, and then moving along to the next problem. The reality, however, is that natural hazards are anything but static; they are dynamic and complex. The nature and severity of extreme events change over time, and climate change will only exacerbate the unpredictability of natural events.

Rather than addressing natural hazards using traditional linear problem solving methods (i.e. addressing problems on a case-by-case basis as they occur), municipalities should employ a risk management approach to adapt to climate change impacts. Risk management helps municipalities incorporate uncertainty into their planning processes, allowing them to develop tools to address specific impacts caused by climate change, enabling continuous monitoring of solutions, fostering a culture of climate change adaptation throughout the municipality.

Exercising dynamic risk management in municipalities is especially important because the impacts of climate change are ever-changing, and many uncertainties exist. Failure to adequately address or acknowledge the wide array of both known and unknown impacts in problem solving can create other even more wicked problems.

It is important to understand the nature of the problem at hand, and to integrate the unknowns into management processes. The incorporation of risk management will ultimately increase municipal resilience to natural hazards, and will ultimately increase the sustainability of a city.

Adaptation and Risk Management

Risk management employs a systematic process of establishing the context for the treatment of risks, then identifying, analyzing, evaluating, treating, monitoring, and communicating risks associated with an organization's activities, functions, and processes (see Figure 6-1). Rather than practised as a separate activity, risk management is best served when embedded in an organization's culture.[23]

Figure 6-1

Risk Management Process

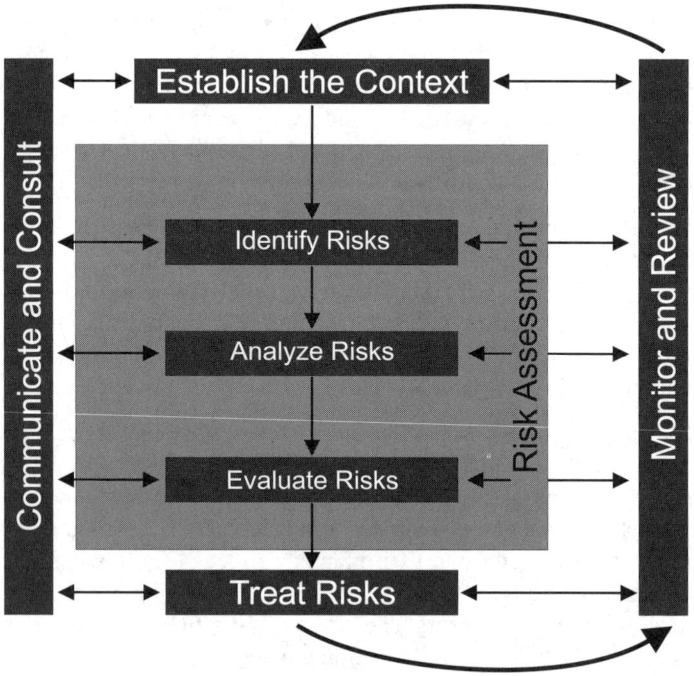

```
                    ┌─────────────────────────┐
                    ▼                         │
    ←→  Establish the Context  ←→
    ┌──────────────────────────────────┐
    │         Identify Risks       ←─  │
    │              ↓                   │
    │         Analyze Risks        ←─  │  Risk Assessment
    │              ↓                   │
    │         Evaluate Risks       ←─  │
    └──────────────────────────────────┘
                    ↓
    ←→       Treat Risks        ←→
    │                              │
Communicate and Consult     Monitor and Review
```

Standards Australia. (2004). *Australian/New Zealand Standard in Risk Management* (AS/NZS 4360:2004).

To promote a culture of risk management, the process should be initiated by a high-level political leader (i.e. the mayor). Political leadership generates media attention, and helps public awareness of climate change adaptation strategies adopted by the municipality. High-level political leadership also promotes a coordinated approach to risk management planning that spans the traditional "silos" of municipal planning.

Effective climate change adaptation requires action from homeowners, business people, and other residents. Thus, the public needs to be involved in the adaptation and risk management decision-making process. Involvement of the public allows for the further identification of risks and vulnerabilities, and fosters a culture of hazards awareness and adaptation at the individual level.

By using the risk management framework, municipalities can identify, analyze, and evaluate the risks associated with climate change. Once analyzed, the appropriate and available tools for adaptation should be identified, developed, and used within the municipality. Many management tools are available, and may include building code by-laws, land use planning, public education, and public health strategies.

To effectively adapt to climate change impacts, the public must be informed and made aware of the potential for increasing damages. Tools, including public consultation meetings and workshops, provide the public with the knowledge and means to protect themselves and their property from increasing natural hazards. The public must also be aware that some of the negative impacts of climate change will be unavoidable.

Enhancing Resilience and Sustainability

In the study of natural hazards, sustainability means that a locality is able to tolerate and effectively recover from damage, diminished productivity, and reduced quality of life inflicted by an extreme event, without significant outside help. When a disaster strikes, a resilient city incurs less damage to its infrastructure, buildings, and social and economic systems.

Thus, the resilient community requires fewer resources to recover from damages, and is able to return to normal functions far more quickly than a community that has not incorporated resilience into its planning models. The fewer resources wasted on recovering from disasters, the more resources available for development and projects that will further increase the community's sustainability.

Resilience is the ability of a municipality to overcome adversity and to recover quickly after a damaging event. As an analogy, engineers incorporate resilience into planning and design. When designing a bridge, an engineer will have a realistic expectation of the maximum weight that the bridge may need to support at any given time (i.e. a "theoretical limit"). However, they will always include a safety margin – at some time in the future, the bridge may be tested with more weight than the engineer would realistically have expected at the time of design. Due to safety margins, the bridge will withstand loads that were unforeseen when it was designed – the bridge is resilient.

Similarly, the municipality should account for future climate scenarios that exceed the capacity of its existing infrastructure. Just as the engineer increases the safety margins when designing the bridge, the municipal planner must increase the safety margins of policies to account for the uncertainty caused by climate change.

Risk management helps a community develop "safety margins" by developing tools and promoting a culture of hazards awareness. Risk management assists municipalities in addressing climate-related hazards before they become problems.

Current Municipal Actions

Considering the significant history of viewing the environment and addressing hazards as static and predictable, it may take some time before climate change adaptation is fully integrated into any municipality's operations. However, some municipalities have made progress in a number of the risk management processes. As well, several municipalities, although not directly addressing climate change adaptation, have incorporated processes to foster a culture of hazards awareness, which can be used for climate change risk management.

The Halifax Regional Municipality formally launched its Climate Sustainable Mitigation and Adaptation Risk Toolkit (Climate-SMART) in March 2004. Development of the program involved partnerships between the municipality and local businesses involved in climate change adaptation, under the ClimAdapt network. Tools to be produced by the program include:

- a risk management tool;
- a community-based vulnerability assessment and risk management tool;
- a cost-benefit assessment tool;
- an environmental impact assessment tool; and
- a communications and outreach tool.

The purpose of the program is to develop a fully integrated planning approach that addresses the impacts of climate change. The program includes guidebooks for individuals, businesses, and communities on how to reduce their emissions, including emission reduction targets, and tools for adapting to the impacts of climate change.

As the Climate-SMART program develops, the aim is to incorporate climate change impact adaptation into regional planning, including effective land use planning to reduce climate risks. Several tools have been developed, and are currently available on the ClimAdapt website at <www.climadapt.com>.

Toronto's Clean Air Partnership (CAP), funded through the Toronto Atmospheric Fund, is working to reduce greenhouse gas emissions in Toronto, as well as to develop strategies for adapting to climate change impacts. As of summer 2007, CAP has been working on a four-part strategy, including an assessment of potential impacts on the city; an assessment of what other communities are doing to adapt; a decision-making workshop; and a report on adaptation strategies. The program has made progress in the risk identification and evaluation stages of the risk management process, and published a report outlining some of the potential impacts of climate change. The report includes outlines of how climate change might affect specific municipal sectors, including water and wastewater, health, buildings, vulnerable populations, and the economy.

While not directly addressing adaptation to climate change, the cities of Peterborough and Edmonton have made progress in creating a culture of hazards awareness. Following flood events, the cities of Edmonton and Peterborough conducted public meetings designed to inform residents of flood mitigation actions taken and actions that residents can take to reduce their risk of sustaining future damages. The meetings were also designed to allow the public to provide input into municipal flood mitigation decisions. In Peterborough, residents were asked to comment on the nature of flooding that occurred, including damage levels and water depths. Residents were also asked to provide input into their priorities for flood reduction. Through this, residents informed the city that their priority was on the reduction of sewer backup, over the reduction of overland flooding.

Public meetings following flood events in Edmonton provided that city's residents with practical information on how to protect their homes from future flooding, and shared details about the actions the city was taking to reduce flood occurrences. Residents were also asked to provide their input on flood reduction measures, such as the location and nature of stormwater management ponds.

Although public meetings in Peterborough and Edmonton did not directly relate to adaptation to the risks of climate change, they serve as a model of how a municipality can engage stakeholders (residents) and foster a culture of risk management within the public.

Novel climate change adaptation strategies are also being developed internationally. For example, an adaptation strategy for King County, Washington is currently under development. The King County initiative is particularly progressive, and includes a goal of an 80 percent reduction in

GHG emissions by 2050. The plan also outlines a framework for continuous monitoring and revision of the climate change mitigation and adaptation program. Considering the development of unforeseen changes in the climate, and the development of new knowledge related to climate science, the plan outlines strategies for adjusting the program based on new knowledge and unforeseen circumstances.

Coordination is a key component of the plan, as it incorporates efforts from the departments of development and environmental services, transportation, natural resources and parks, and public health, among others. Partnerships have also been established with the Climate Impacts Group at the University of Washington, and the International Council for Local Environmental Initiatives (ICLEI). A climate change guidebook for local, regional, and state governments has been produced in partnership with ICLEI, and serves as a tool to assist municipalities in developing resilience to the impacts of climate change. This guidebook can be downloaded from the ICLEI website at:<www.iclei.org/documents/USA/download/0709climateGUIDEweb.pdf>

The cities of London (UK) and New York (US) are also developing climate change adaptation strategies and toolkits. These cities and other municipalities around the world are taking advantage of public/private partnerships for funding and expertise. Canadian municipalities should leverage available international experience, knowledge, and resources, to increase their capacity for climate change adaptation.

Conclusion

Canadian municipalities will need to infuse risk management processes into their organizational cultures in anticipation of an increase in the occurrence and severity of natural hazards. Effectively building resilient communities requires that municipal governments abandon the traditional model of studying the problem, implementing a solution, and moving on to the next problem. There are no simple solutions, because the variables and uncertainties are ever-changing. As a result, the best practices and impact management techniques remain in their infancy. A good first step is to address the impacts through risk management, facilitating an effective approach to climate change adaptation.

The environment and natural hazards have been traditionally viewed as static entities; thus, it may take some time for communities to fully embrace the risk management process. However, Canadian municipalities can take advantage of adaptation tools and methodologies already available to begin the management process now. Incorporating these will allow for a smoother transition from the traditional hazards management approach to one of risk management – moving us toward more resilient communities.

Chapter Seven

Education for Sustainable Development

Lyle A. M. Benko

How do we engage citizens on climate change, so that they are part of the solution? How do we move them from simply being aware of what they should be doing (i.e. reducing GHGs), to actually doing it? These are important questions, because just as governments have critical roles to play in response to climate change, so too do individuals.

This chapter examines the important role of education in engaging citizens on climate change and sustainability. Education provides an opportunity for communities to become more than just aware, but to understand and become committed to action. It is through the collective actions of individuals that we will address climate change and achieve sustainability.

Moving from Science and Policy to Understanding and Action

Government bodies are responsible for creating policies and principles to help make our societies and communities sustainable. In the next chapter, Susan Gardner discusses how local governments, for example, can tackle climate change planning through the Integrated Community Sustainability Planning process. Indeed, much of this book discusses different ways that local governments are addressing climate change.

But, climate change is such a complex issue. How can those long scientific reports be translated meaningfully to individuals and communities, so they become aware of and understand their individual role in society? What determines how people move from simple awareness to real actions? Is it based on their understanding, concern, and commitment to helping make their community a more sustainable living environment? How can we most effectively move people into action, other than by simply advertising new policies, regulations, or incentives?

One thing is clear: it is critical for politicians to understand the science and facts of climate change. But, just as important is the need to help the average citizen discern the politics from the facts, so they can understand how the issue impacts on them and their communities.

This has not been lost on the City of Regina, which has made a real effort to engage citizens on climate change. With the help of its Green Ribbon Community Climate Change Advisory Committee, the city has learned that, with a sustained effort, it is possible to move beyond political debates to tangible actions by citizens.

Several years ago, city council appointed a team of leaders in the community to address the reduction of GHGs by the citizens of Regina. This advisory committee developed an action plan and a subsequent *Green Book: $Mart Ways To Save*. The focus of the publication was on ways and means of saving money and also reducing GHG emissions. This publication has been distributed throughout the community and is currently going into its third revision of publication. The articles and ideas in the book focus on success stories of local residents that have already moved from awareness to action.

Awareness to Action

A few years ago, when the provincial Climate Change Hubs were developed across Canada by the federal government, Saskatchewan developed its own hub and a second entity called Climate Change Education Saskatchewan (CCES). Over several years, CCES developed very specific education programs that were incorporated into the Kindergarten to Grade 12 curriculum. Teachers were involved in developing materials that were specific to Saskatchewan and were approved and integrated into the curricula. Many teacher workshops were held throughout each of the seven regions of the province, so that educators could be introduced to the materials, and most importantly, become educated with the scientific facts and issues related to climate change. Teachers were also encouraged to become "models" for their students and their communities, with the locally developed curricula for Saskatchewan.

CCES (and other similar organizations) continue to provide support for educational programs. Now, the Government of Saskatchewan just released its *Green Strategy: For a Green and Prosperous Economy* (2007), that focuses on informed and engaged citizens and communities. It recognizes the need to educate the public about the environment, encouraging individual commitment to action to reduce environmental impact. In Saskatchewan, the move from policy to practical changes is being supported by all orders of government.

At the same time as national level support for educational and public outreach programs was being revised, a new approach to teacher workshops was being developed at the University of Regina Faculty of Education. This approach involved workshops for the pre-intern teachers before they went out to teach in their final year of internship. The workshops focussed on educating these young teachers about the science and issues of climate change, and also on helping them become aware of the much-needed support materials and curricular connections for their future classrooms. The workshops were delivered by post-interns that were in their last semester completing their degree. These future teachers were exposed to a model that moved them through the following stages: awareness, understanding, concern, commitment, and action.

Each stage was designed for the teacher to become aware of their own knowledge, or lack of knowledge, about climate change. They were then provided with materials and resources to help them understand the difference between scientific facts and the political debates. This led to exploration of the concern they might have as individuals, and equally importantly as educators, in helping society understand the same information, as well as their role in helping resolve the issues. Then, they began to explore their own personal and professional commitment to develop how they would achieve solutions to the concerns that were raised earlier. Finally, they were to develop action plans for their future role as an educator and community member.

Workshop participants indicated in their workshop evaluations that they had moved further in their understanding and commitment than if they had previously just seen a brochure or advertising campaign about climate change. They also indicated that they had become more committed to doing something in their own personal lives that would model actions for their own students. Through follow-up conversations and interviews with some of these teachers, they identified their own personal action plans that were put into place for themselves, their schools, and their communities to address the issues of climate change.

Education for Sustainable Development

Sustainability entails a balance between the needs of the economy, environment, and society. This evolving concept is very different for individual communities. It is therefore incumbent on local communities to determine what is the right balance within their boundaries.

Education can provide an opportunity for the local community to become more than just aware of the policies, to understand and become committed to actions for issues such as climate change. This may bring about the balance and sustain the community into the future. It can happen by using local examples that are already occurring within the community, showing the connection between the economy, environment, and the society in that particular community.

The role of formal education in helping communities was officially recognized in January 2005 when the United Nations (UN) established the Declaration for the Decade on Education for Sustainable Development (2005-2014). This followed on several years of supporting research through the United Nations University. The declaration recognized the need and importance of education in bringing about change in global issues, and the need for these changes to begin at the local "grassroots" level. The declaration also recognized the need for development of education programs to also assist the informal (media) and non-formal (corporate) sectors of the community to understand their role and significance.

United Nations Regional Centres of Expertise

Moving from a locally developed educational program on climate change issues to a global perspective also became evident in January 2005, with the establishment of the United Nations Regional Centres of Expertise on Education for Sustainable Development (RCEs). On March 1, 2007, Saskatchewan became the second UN Regional Centre of Expertise (SASKRCE). Toronto was the first RCE in North America, and Sudbury also recently became recognized. There are several others in North America working towards their designation by the UN. Currently, there are over 35 RCEs around the world with the UN designation.

The focus of these RCEs is to achieve the goals of the UN Decade of Education for Sustainable Development, by translating its global objectives into the context of the local community. This is very evident in Saskatchewan, where climate change was identified as one of the six main themes to be researched and subsequent actions developed. The main channels of achieving these goals include:

- access to quality basic education;
- reorienting existing education;
- public awareness and understanding; and
- training programs for all sectors.

Local to Global Action

Moving from the local perspective to a global view helps build capacity and collaboration through shared expertise. With the establishment of SASKRCE, several local municipalities such as Regina, Craik, and Saskatoon collaborated with organizations like Saskatchewan Outdoor and Environmental Education Association, and institutions such as the

University of Regina, University of Saskatchewan, Saskatchewan Institute of Applied Science and Technology, as well as provincial government bodies such as Saskatchewan Environment and SASKPOWER. Nearly 30 entities documented their contributions and existing capacity to support the United Nations application for an RCE. It documented the strong collaboration, capacity and existing expertise that already existed to deal with climate change and other sustainability issues.

The six Education for Sustainable Development issues identified by the Saskatchewan RCE included:

1. climate change
2. health
3. farming and local food production, consumption, and waste minimization
4. reconnecting to natural prairie ecosystems
5. supporting and bridging cultures for sustainable living and community building
6. sustainable infrastructure, including water and energy.

With these six issues identified in the region, SASKRCE will focus on two cross-cutting areas of research and engagement by individuals:

▶ sustaining rural communities; and

▶ developing regionally appropriate educational approaches for Education for Sustainable Development.

Since the March 1, 2007 designation of SASKRCE by the United Nations, work has already begun to develop action plans and research projects based on the goals and issues identified. This has been recently enhanced by the formation of the Saskatchewan Education for Sustainable Development Working Group. This provincial working group has joined other provincial working groups to form a national network to focus on the role of formal education to assist communities in a better understanding of their own sustainability.

Conclusion

The important role that education plays in moving people from awareness to action needs to be recognized as a major way of helping individuals to change their behaviour and become engaged in resolving issues in their community. The role of municipal, provincial, and federal policy makers is to provide leadership, but it will be the individual actions of people in their communities that will bring about our sustainable future, and the necessary balance between the needs of the economy, environment, and society.

It appears that individuals in society have become much more aware of the topic and issues related to climate change and its impact on the global scene. The challenge now remains for local municipalities to show leadership in modeling sustainability through their policies, decision making, and the technological advancements they make in their corporate operations. The corporate entity of the City of Regina has already begun to show this leadership to its citizens in recent energy efficient changes and upgrades it has completed for city-owned buildings. Further actions are planned to act as a model for citizens to also become engaged in similar actions to reduce greenhouse gas emissions.

The manner in which the Green Ribbon Committee has developed and moved its Action Plan forward, with the support of council, is an excellent model that may provide further research and data for other UN Regional Centres of Expertise in Canada and elsewhere in the world.

With the development of the United Nations Decade of Education for Sustainable Development and the Regional Centres of Expertise, the opportunity will be greater for individuals in their communities to become involved with research and engagement on key issues at a global level. The term: "Thinking Globally and Acting Locally" will be realized through the important vehicle of Education for Sustainable Development.

Additional Reading and Resources

United Nations Regional Centre of Expertise: <www.ias.unu.edu/research/details.cfm/ArticleId/466/search/yes>

City of Regina Green Ribbon Climate Change Advisory Committee: <www.regina.ca/content/info_services/climate/greenribbon/commitee.shtml>

Saskatchewan's *Green Strategy: For a Green and Prosperous Economy*: <www.saskatchewan.ca/green>

Saskatchewan Regional Centre of Expertise <www.saskrce.ca>

Saskatchewan Education for Sustainable Development Working Group: <www.saskesd.ca>

Climate Change Saskatchewan: <www.climatechangesask.com>

Chapter Eight

Integrated Community Sustainability Planning

A Process for Addressing Climate Change at the Local Level

Susan M. Gardner

"Sustainability" has truly emerged as one of the most powerful imperatives of this generation, with issues like the climate change crisis underscoring for Canadians the interconnectedness of the natural world with the economic, social and cultural aspects of our society. This has required a re-evaluation of how the business of public policy making is conducted in every order of government across Canada. The path to achieving sustainability is clearly not a simple one, and requires a breaking-down of the silos that isolate individual policy areas, and the embracing of a more integrated approach.

Integrated Community Sustainability Plans (ICSPs) are one of the key processes currently available to Canadian municipalities in forging that path, to deal with the many challenges facing them today. Because of the heightened public awareness around climate change, ICSPs also present municipalities with an ideal opportunity to address this important issue, both in terms of adaptation and mitigation strategies, as part of a holistic approach to planning for the community's future. ICSPs can also be instrumental in helping municipalities to move beyond rhetoric to action.

Understanding ICSPs

Municipalities across Canada are required to develop Integrated Community Sustainability Plans as part of the gas tax agreements signed with the federal government. The gas tax agreements in each province and territory require the signatories to commit to develop ICSPs at the local level, and allow for municipalities to allocate a portion of their gas tax funds to development of the plan. In addition, as discussed in Chapter 16, grants are available under FCM's Green Municipal Fund for the development of sustainable community plans that demonstrate an integrated systems approach to addressing community-wide energy and environmental management objectives.

As a condition of the agreements in every jurisdiction across the country, municipalities that receive gas tax funds will be required to demonstrate through their existing planning instruments and processes, or through the creation of a new planning document, the following criteria:

- ▶ coordinated approach to community sustainability (eg. linkages of various plans, and planning and financial tools that contribute to sustainability objectives);
- ▶ integration of social, cultural, environmental, and economic sustainability objectives into community planning;
- ▶ collaboration with other municipalities, where appropriate, to achieve sustainability objectives;
- ▶ engagement of residents in determining a long-term vision for the municipality.

Because climate change is one of those issues that cross-cuts all four pillars of sustainability – social, cultural, environmental, and economic – the requirement for Integrated Community Sustainability Plans presents communities with the perfect opportunity to reflect on climate change as a significant factor in their environmental scans, and to develop comprehensive strategies and plans for action.

Climate change is known to have far-reaching impacts within each of the pillars of sustainability. As one of the most urgent concerns currently facing governments at every level – and one on which the public in many communities is already strongly engaged – climate change can have an important role as a catalyzing issue to drive the development of ICSPs, and to encourage holistic policies that take all aspects of sustainability into consideration.

As every municipality is unique, and legislated planning requirements also vary considerably across jurisdictions, there is no single roadmap for the development of an ICSP. Provincial governments and municipal associations in each jurisdiction have been working to create tools that will help municipalities develop or adapt plans that will meet the criteria.

Some municipalities, like Whistler, BC, have used The Natural Step Framework – a proven systems approach for dialogue and decision making – to develop sustainability plans for their communities. The Natural Step Canada has now developed a process for assisting cities and communities both with the development of new ICSPs, and with building upon existing municipal planning instruments and processes. In addition, Drs. Chris Ling, Ann Dale, and Kevin Hanna of Royal Roads University have recently developed an "Integrated Community Sustainability Planning Tool," available from The Natural Step Canada website at <www.naturalstep.ca/scp/sustainablecommunities.html>.

Other initiatives, such as BC Healthy Communities (BCHC), can also work in concert with the ICSP process. Launched in the Fall of 2005 with funding from the BC Ministry of Health, BCHC is part of the international Healthy Cities/Healthy Communities movement. The initiative supports communities and community groups that are taking a holistic and integrated approach to increasing the health, well-being and healthy development in their communities, and provides resources to help communities engage diverse sectors to this end. An Integral Capacity Building Framework is used, with a four-step process designed to promote capacity building activities in the following areas: learning; engagement; expansion of assets (eg. knowledge, relationships, resources, partnerships, activities); and collaboration in defining a path to reach a chosen future.

It is important to recognize that tools like The Natural Step and BCHC's Integral Capacity Building Framework are not mutually exclusive, but can work together to achieve a multitude of goals and mandates, engaging the public in that process. The public engagement component is critical to changing behaviours and attitudes, and to successfully charting a shared course for the future. Also, the requirements for the development of ICSPs under the gas tax agreements specify that municipalities must demonstrate that they have engaged residents in determining a long-term vision. For this process of engagement to be meaningful and effective, community leaders will need to start by defining what "sustainability" is – as explored further below – and develop language that gives people in the community a shared understanding of the concept; programs like The Natural Step and BCHC can help communities to initiate this local dialogue.

Defining Sustainability

As intuitive as the terminology might seem – because it is now widely used throughout the municipal sector – "sustainability" is a relatively new concept, and one that is not widely understood (or commonly understood) by most Canadians.

In a 2006 study by James Hoggan and Associates, 53 percent of Canadians said they had never heard of the term "sustainability." In addition, seven in 10 were unable to define "sustainability." However, once the term was defined, over 80 percent of Canadians rated sustainability as a top or high priority national goal. So, the definition of sustainability is a key piece.

In 1987, the United Nations Commission on Environment and Development (the Bruntland Commission) drew attention for the first time in a major way to the fact that economic development often leads to a deterioration, not an improvement, in the quality of people's lives.

The Commission therefore called for "... a form of sustainable development which meets the needs of the present without compromising the ability of future generations to meet their own needs" – the definition of "sustainability" accepted by most practitioners today.

In this, there is recognition that we need to think beyond economic development, and take into consideration the impacts of our activities on the environment, and also in a social sense, in terms of quality of life; there is a sense that we need to act in ways that enhance our quality of life today, but also protect it for future generations.

The Melbourne Principles

A major piece of the sustainability framework emerged in April 2002, when an international group of experts – including representation from Canada – came together to develop the Melbourne Principles for Sustainable Cities. The 10 principles were developed to assist cities that wish to achieve the sustainable development objective enunciated by Bruntland:

1. Provide a long-term vision for cities based on: sustainability; intergenerational, social, economic, and political equity; and their individuality.
2. Achieve long-term economic and social security.
3. Recognize the intrinsic value of biodiversity and natural ecosystems, and protect and restore them.
4. Enable communities to minimize their ecological footprint.
5. Build on the characteristics of ecosystems in the development and nurturing of healthy and sustainable cities.
6. Recognize and build on the distinctive characteristics of cities, including their human and cultural values, history, and natural systems.
7. Empower people and foster participation.
8. Expand and enable cooperative networks to work towards a common, sustainable future.

9. Promote sustainable production and consumption, through appropriate use of environmentally sound technologies and effective demand management.
10. Enable continual improvement, based on accountability, transparency and good governance.

The Melbourne Principles are intended as a guide to thinking and a strategic framework for action, where citizens and decision makers can work together to transform their communities. This aspect of civic engagement – of having citizens and decision makers working together towards sustainable ends – is a central tenet of the new sustainability paradigm, and is essential to ensuring the success of today's efforts to address climate change issues. However, while the principles provide a simple set of statements on how a sustainable city would function, they don't provide a road map. To operationalize the principles, other tools or processes – like ICSPs – are necessary.

Nonetheless, a discussion of the Melbourne Principles can be very useful to get people thinking and talking about the kind of community they want for the future – the values they hold today, current threats to their quality of life, and the kinds of initiatives they would like to implement and participate in to address those issues.

The Pillars Approach

Since Melbourne, there has been a growing dialogue about the "pillars" of sustainability and how they relate to sustainable communities. Many plans refer to a "triple bottom line" model, which involves evaluating public policy in terms of being economically, socially, and environmentally sustainable. A more comprehensive approach is evolving, though, that includes a consideration of "cultural" sustainability as well.

The last element – cultural sustainability – is widely understood to include things like arts, history, and heritage – and it does. But its scope is actually much greater than that. Culture is perhaps most accurately described as our community value system.

It's been suggested that without considering our culture as in integral part of the sustainability equation – by determining and defining community values – we are unable to understand and address the sustainability issues around the other three pillars. The bottom line is that if we don't know what, in fact, the community values, we don't know what is worth sustaining.

Once those values are defined, ultimately, the challenge for communities becomes "How does this all fit together?" Answering that challenge is ultimately the goal of ICSPs, as they link up objectives from across the pillars, and involve the entire community in developing a plan for the future.

Piecing It All Together: A Paradigm for the Future

Considering the "big picture view" offered by sustainability theory and the consequent emergence of ICSPs, it is evident that there has been an evolution of policy development practice – from viewing policy issues in isolation, to considering all of them as interconnected – an ecosystem, where changes contemplated for one area must take into consideration the impacts on the other pillars. Where even one is weakened, the whole structure is compromised.

In the past, communities have often developed policy pieces in isolation, around the individual pillars, resulting in environmental policies; economic development strategies; social policies; and cultural policies. Those discrete or distinct kinds of policies are important, and all have a role on the local government agenda. But, distinct policies are not enough.

Perhaps more than pillars, these four aspects of sustainability must be viewed as a multi-faceted lens, through which all policy must be filtered (see Figure 8-1). This lens provides a new way of thinking about local decisions.

Figure 8-1

Conclusion

Sustainability is not just about surviving; it's about thriving. For that to happen in Canadian communities requires a paradigm shift, and a re-evaluation of how the business of public policy making is conducted at every level. Most importantly, it requires local decision makers to actively and authentically involve the community in defining how that re-evaluation will occur, and in developing a vision for the future. The tools and resources are now in place to allow that happen – what's needed now, are commitment and action.

Climate change is a global issue, with local impacts that will vary in every community – as will the strategies necessary to deal with it. Climate change is thus an ideal issue to address through the Integrated Community Sustainability Planning process. The challenge will be for community leaders to literally "think globally and act locally," with customized ICSPs that will meet the current needs of the local community, and also protect what that community most values for its future generations.

Additional Reading and Resources

ICLEI—Local Governments for Sustainability, <www.iclei.org>.

The Fourth Pillar of Sustainability: Culture's essential role in public planning, Jon Hawkes (2001), Common Ground Publishing Pty. Ltd. in association with the Cultural Development Network (Vic), Australia.

The Melbourne Principles for Sustainable Cities, United Nations Environment Programme, Division of Technology, Industry and Economics <www.unep.or.jp/ietc/Focus/Melbourne Principles/English.pdf>.

The Natural Step Framework for Sustainability, The Natural Step Canada, <www.naturalstep.ca/framework.html>.

The Sustainability Revolution: Portrait of a Paradigm Shift, Andres R. Edwards (2005), New Society Publishers, Canada.

World Environmental Accords, <www.wed2005.org/pdfs/Accords_11x17.pdf>.

BC Healthy Communities, <www.bchealthycommunities.ca>.

Chapter Nine

Communities of Tomorrow

Edward Willett

Collaborate. Innovate. Thrive. These are words for everyone to live by, from individuals to businesses to governments; but, they are particularly important concepts for municipalities looking ahead at a future made uncertain by climate change.

Of course, the status quo has never really been an option for municipalities, no matter how much they might wish it. Their size and substantial populations alone ensure that they are always in a state of flux. Neighbourhoods and downtowns evolve, new industries sprout and old industries expire, and external factors – ranging from federal tax policies to oil prices to, yes, weather and climate – have their impact.

The challenge for municipalities, then, is to evolve, and to keep on evolving. That means finding new ways of doing things. That's the idea behind the unique Communities of Tomorrow partnership established in Regina in 2003.

Birth of Communities of Tomorrow

Maintaining civil infrastructure is a major challenge for Canada, with approximately 80 percent of the $12 to $15 billion a year spent by municipalities on infrastructure, going to repairs and renewals. One way to meet this challenge is the development of sustainable infrastructure, commonly defined as "the design, construction, planning and maintenance of infrastructure that meets the needs of the present without compromising the ability of future generations to meet their own needs."

Late in 1999, technology advisors from the National Research Council (NRC) met with City of Regina officials to discuss a possible demonstration project related to sustainable infrastructure research. As those discussions progressed, interest grew in expanding that concept into a Regina-based technology cluster that would focus on sustainable infrastructure.

In 2003, the NRC unveiled plans for a Centre for Sustainable Infrastructure Research in Regina. The centre received $10 million from the federal government to spearhead its development, while the city, the province, the University of Regina, and Western Economic Diversification Canada each pledged $5 million over five years – a combined initial investment of $30 million. Communities of Tomorrow's stated vision is to "facilitate the Regina-based cluster of research, municipal, and industry partners who develop and market world-class products, approaches, and technologies for sustainable and environmentally responsible infrastructure that supports healthy, prosperous, and viable communities."

While there's no specific mention of climate change in the vision statement, or in the mission statement or strategic goals set out in Communities of Tomorrow publications, there is one word that comes up over and over again: "sustainable." However, we can't have a serious discussion about being sustainable without accounting for climate change.

Clustered and Interconnected

The goal of Communities of Tomorrow is to grow in Regina a cluster of interconnected companies around a nucleus of research and development facilities provided by the National Research Council, the University of Regina, and Regina businesses, all working toward sustainability. This cluster, it is hoped, will become a "hotbed" of investment and technology transfer, both for existing firms and for new spin-off companies. The objective is to turn Communities of Tomorrow into a global leader in sustainable urban infrastructure.

The City of Regina's involvement is key to this goal. It is serving as a living laboratory, testing and demonstrating newly developed products, technologies, and approaches. At the same time, it is providing the other members of the Communities of Tomorrow cluster with valuable information about the infrastructure and management needs and challenges of a municipality. Regina is, in other words, a real-world "test bed" for what might otherwise remain nothing more than good ideas.

As the NRC puts it, "The initiative will help the City of Regina achieve cost-effective, community-based actions to meet its existing and future infrastructure challenges in sustainable ways, and to become a national centre for environmental infrastructure management research and innovation. Other communities in Canada and elsewhere are expected to benefit by adopting systems proven in and modelled on Regina."

The NRC has collaborated with other partnerships across Canada to build technology clusters; but, Communities of Tomorrow is unique in that it has a direct source of project funds under its own control. In its relatively short existence, it has already funded a number of projects at various levels, some of which show promise for both mitigation of and adaptation to climate change.

That's important because if Regina – or any other municipality – is to be sustainable into the future, it must figure out how to best manage climate change.

Regina: Land Locked and At Risk

Regina may be particularly vulnerable to climate change, says Dave Sauchyn, Research Coordinator for the Prairie Adaptation Research Collaborative. The projected increase in mean temperatures is "profound" and would alter Saskatchewan enormously. He points out that regions to the south with average temperatures three to five degrees warmer are radically different in terms of ecosystems, land use and crops grown.

The largest projected change in southern Saskatchewan's climate will be the distribution of water. The most serious water related impacts will be felt in winter and spring. Water has been identified as an area on which researchers should focus. The potential impact of drier summers on agriculture is obvious, and may well offset the concurrent advantage of a longer growing season. As well, the warmer atmosphere will be able to store more water, which will mean less water on the ground, on the surface and in the soil. This capacity for the atmosphere to store more water will result in less water on the ground, and more in the air. The net effect is less surface and soil water.

Sauchyn says that, although southern Saskatchewan will probably encounter some unusually wet years, in general the very dry years will be more common. That points to a more extreme and variable climate, with worse droughts than the region has ever seen interspersed with occasional years that are wetter than the region has ever seen.

Land-locked (one of only two land-locked cities in North America) and with its location determined more by chance and 19th-century politics than because of any particular geographic advantages, Regina is a bit like a space station: an artificial environment surrounded by a natural environment that is not always hospitable. The risk of water shortages and droughts is a major cause for concern.

All cities are artificial, to a greater or lesser extent. In the face of climate change, this artificial nature can actually be an advantage. Natural systems, as Sauchyn points out, can only adapt to climate change after the fact. As temperature and water levels change, existing species die out or move out, while new species establish themselves. But, artificial systems like Regina, or any other city, are *built* – and, with sufficient political will, financial resources, and technical know-how, can be *rebuilt* in anticipation of change, rather than in reaction to it.

Innovation for Tomorrow

Communities of Tomorrow welcomes ideas and suggestions from anyone, with more than 100 collected from a variety of sources within the City of Regina. Among the broad areas of interest identified: water chemistry and microbiology; wastewater treatment; roadway construction and operations; structural materials; sidewalk, pathway, and driveway construction and operations; electrical and electronics; construction methodology; and urban forestry and horticulture.

Funding that is available from Communities of Tomorrow for approved projects ranges from $15,000 for short-term exploratory research to more than $150,000 for larger-scale infrastructure research programs. Communities of Tomorrow also helps researchers to access research capabilities and equipment from its research partners, and to test their technologies within the City of Regina.

One project receiving Communities of Tomorrow funds that holds promise for both mitigation and adaptation is the Factor 9 Home. The project's goal is to design, construct and monitor a new detached house in Regina that will use 90 percent less fossil fuel energy (and thus produce only one tenth as much greenhouse gas) as a conventional house, through conservation and use of solar heating.

At the same time, the house will use only half as much water as a typical house, thanks to low-flow toilets and showerheads, a low-flow dishwasher and clothes washer, landscaping that requires little water, and the collection of water from the roof for toilets and exterior use. These are adaptations for a future climate where water supplies are less reliable.

If reservoirs run low during periods of extended drought, or floods result in an increased flow of agricultural chemicals or other pollutants into municipal water supplies, improved methods of monitoring and maintaining water quality will be important elements of an effective response to climate change. In keeping with its initial research priorities, Communities of Tomorrow has funded several projects that focus on water quality. One example: Remote On-line Monitoring of Water Quality in Water Distribution Systems, a project aimed at developing a method of conducting real-time, remote monitoring and evaluation of water quality in small to medium-sized communities and water utilities.

Another example has been a water treatment project that uses a new electromagnetic and optical water treatment process to eliminate pathogens. The process will be simple, energy efficient, and economical, and can be retrofitted to any drinking water treatment system. When planning for a sustainable future, all systems and infrastructure must be made as efficient and innovative as possible.

Finally, if a technology is proven to be commercially viable, Communities of Tomorrow links researchers with public and private partners. Partnerships like this have been integral in the development of capacity building programs that outlast electoral cycles. Innovative ideas and sustainable infrastructure that can contribute to economic growth are attractive incentives for successful partnerships. By providing funding to research and industry, the risk to business is reduced, and the competitiveness of new commercial products on the global market is increased.

Conclusion

Cities and municipalities can proactively protect their climate by anticipating and managing threats before they arrive. Says Sauchyn, "Cities have at their disposal the resources to deal with climate change. But that capacity to adapt is only that, it's only capacity. You need the will to mobilize those resources."

"One of the key barriers to adaptation currently is a lack of leadership and a lack of public policy to coordinate the process. There is very much that individuals are doing about climate change. But, that's spontaneous and ad hoc, and just based on their own motivations and initiatives. What we need is a much more coordinated and facilitated effort to deal with climate change. That's going to require political leadership and some policies and programs."

Communities of Tomorrow is one such coordinated and facilitated effort. It's one that promises to help point the way for Regina – and for other municipalities that draw on the knowledge and technology developed through the projects it helps fund – to mitigate and adapt to climate change, and to evolve sustainably into the future in many other ways.

In other (much more succinct) words, Communities of Tomorrow helps communities to collaborate, innovate, and thrive.

Chapter Ten

Who has the Answers?

A Small Community Perspective

Bill Beamish

In 2006, many coastal communities like the Town of Gibsons, British Columbia were pummelled by extreme weather events with levels of intensity and frequency not previously experienced. Significant damage was done to private and public lands as a result of snow, heavy rains, and wind storms. In December 2006, over 240,000 BC Hydro customers were without power, and more than one million people were affected by a lengthy boil water order from local health authorities. In addition, municipal governments in the Lower Mainland were hit with a $12 million bill for recovery and clean up of 29,000 trees that were lost in winter storms, including 10,000 trees felled in Vancouver's Stanley Park alone.

These adverse weather patterns continued in 2007, with reports in February and March of mud slides closing down the Trans Canada Highway and the CP/CN railway lines, and washing away the banks of rivers, threatening nearby homes and properties. These slides were the result of heavier than average rainfall that has saturated the ground and flooded local rivers. Many local governments throughout British Columbia were prepared for a year of record-level flooding due to heavy snow packs and the predictions of warmer weather. Although rivers peaked below expected levels, hundreds of homes were evacuated and thousands of lives disrupted across the province.

So, as we sit comfortably overlooking a small fleet of commercial fishing boats and pleasure craft moored in the calm, protected harbour of Gibsons (population 4,022, and "Home of the Beachcombers"[24]), we can only wonder about how global warming and climate change will affect our community in years to come. Will these weather patterns continue? And, if they do, what can we do to prepare and protect our communities?

Who Has the Answers?

Residents of Gibsons value their access to the low bank waterfront and harbour area. Owners of expensive waterfront property are planning new residential and commercial development/redevelopment that will, in keeping with the recently completed official community plan by-law, maintain the small town atmosphere of the community.

While observing or experiencing extreme weather events, we are also becoming more aware of and concerned about the potential impacts of climate change. The extensive media coverage of world events, and reports on the findings of recently published scientific studies by the United Nations, Great Britain and others, create further awareness and concern. Perhaps, the most effective and informative influence has come from the film documentary by Al Gore, *An Inconvenient Truth*, which has been widely viewed in theatres, offices, church basements, and private homes, and was awarded the Oscar at the 2007 Academy Awards for best documentary film.

Many questions are being asked about the future of our community, particularly of our lovely waterfront areas: Are the breakwaters high enough to protect the harbour from rising ocean levels? Is the seawall sufficient to keep water from flooding adjacent properties and homes? Are our development standards sufficient to ensure that new buildings will withstand rising water levels or storm surges? Is our community infrastructure safe? What about our aquifer and watersheds? Is the "World's Best Municipal Drinking Water"[25] threatened?

As our residents become more aware and concerned about climate change, they are looking to their locally elected officials and town staff for answers and direction. But where do *we* look?

The unfortunate reality is that small communities like Gibsons haven't the capacity to sort through the myriad of scientific reports and opinions – some of which are contrary – about climate change and its potential to impact our community. Help is needed to sort through public information and scientific reports, so that we can make informed decisions about the future of our community.

Responding to Local Concerns

The Town of Gibsons council is responding to local concerns by engaging the community. Gibsons is also working with the University of British Columbia to identify community vulnerabilities, and to develop and implement adaptive strategies and best practices for Gibsons, and possibly for other coastal communities. In 2007, council initiated a study of the Gibsons harbour and waterfront areas to determine what kind of development is possible and desired by the community. The town is also working to identify risks, and to determine if new development standards are required to ensure that existing and proposed development and community infrastructure are protected from the effects of climate change.

Of paramount concern are: the impacts of rising sea water levels on waterfront development, and on the town's aquifer that provides two-thirds of the community water supply; the need to establish new standards for residential and commercial construction that will adequately deal with increased rainfall and stormwater runoff, rising water levels, and any changes to the groundwater table; protection of harbour infrastructure from storm surges and heavy winds; tree maintenance; and management of public areas and private lands that will be impacted from winds and storm events.

Gibsons has turned to other governments and organizations to provide accurate and timely information and assistance to deal with the issues that climate change presents. Examples of sources consulted and relied on by Gibsons are:

- other orders of government for studies and financial assistance for projects;
- other local and regional governments for model by-laws and policies;
- municipal associations, including:
  - Federation of Canadian Municipalities;
  - Union of British Columbia Municipalities;
  - Association of Vancouver Island and Coastal Communities;
  - Coastal Community Network;
- non-governmental organizations:
  - Sierra Club of Canada;
  - Sunshine Coast Air Quality Society;
- open source information:
  - media – newspapers, radio, television, and magazines,
  - books, trade journals, and scientific reports;
  - Internet;
- consultants;
- universities, colleges, and local schools:

- University of British Columbia, Institute for Resources, Environment and Sustainability;
- Canadian Climate Change Impacts and Adaptation Research Network (C-CIARN).

The Town of Gibsons recognizes the importance of creating community awareness around the need to invest in protective and adaptive measures today, before it is too late to adequately respond to changing environmental conditions.

What Gibsons Has Done and Is Doing

As a community, Gibsons is justly proud of the many initiatives that it has taken to preserve and protect the local environment. Since 2000, council has implemented over 20 programs designed for this purpose. These include initiatives to improve air quality, such as an anti-idling by-law, the prohibition against any open burning, and the Howe Sound – Sea to Sky Air Quality Management Program, an agreement between local industry and other communities. These and other initiatives were achieved while engaging in research partnerships with the University of British Columbia.

Other initiatives include the incorporation of "Smart Growth" principles into the town's official community plan. The "Smart Plan," which incorporates a triple bottom line approach to development, was adopted in 2005. It has since been recognized by the Planning Institute of BC and Smart Growth BC.

In 2006, Gibsons council passed a resolution that acknowledges the reality of climate change and global warming and its potential devastating effects on communities like ours. In so doing, council has also encouraged other local governments and orders of government to make this issue a priority, and to work together to find appropriate solutions.

In 2007, Gibsons council resolved to join the FCM Partners for Climate Protection Program (PCP) and passed the FCM's model resolution on climate change. It also became a signatory to the British Columbia Climate Action Charter, committing its operations to become carbon neutral by 2012, and to measure and report on community GHG emissions as it pursues the goals of becoming a green community.

Gibsons council and staff have also made a conscientious effort to get ahead of the curve on this issue, and are actively pursuing initiatives that will help them to stay informed and better able to respond to or engage the community. Some examples of these initiatives that could be considered by other communities are:

Communicate/dialogue with local community

1. *Discussion sessions with council and staff* – Each month, one lunch hour is set aside for informal showings of environmental or community development programs, followed by a group discussion of the issues and how they relate to our community. A good place to start is the documentary *An Inconvenient Truth*.
2. *Community dialogue* – Monthly two-hour discussions are held with the community to explore issues of concern, and to share information on town initiatives, including work on climate change.
3. *Newsletters and town website* – A quarterly newsletter is planned to provide information to the community and to invite communication with council. Gibsons also provides a very comprehensive website <www.gibsons.ca> on community issues and council business.

Develop relationships with key partners

1. *C-CIARN* – Although the program was terminated in 2007, researchers that were involved in the C-CIARN project provided valuable support and published some very interesting and useful information about the impacts of climate change for local governments.
2. *University of British Columbia* – The university has been very helpful and supportive of the community efforts in Gibsons.
3. *Nearby communities and regional governments* – Gibsons is not the only community that will be affected by climate change. Developing strong working relationships or partnerships with nearby communities and regional governments is important. They can provide advice or assistance in planning and implementing measures in response to impacts.

Exchange information with other communities/orders of government

1. Best management practices
2. *Presentations to intergovernmental meetings and to regional and provincial associations* – It is important that relevant information is shared as widely as possible, so that everyone benefits from the work done elsewhere, and that it contributes to the growing bank of knowledge and experience.

Apply for grants to assist with local projects

1. *Federal and provincial governments* – Local governments do not have the capacity to fund in-depth research projects and studies. Programs are available and will likely grow as federal and provincial governments make climate change a priority. Plan now and be ready to act when the opportunity becomes available.

Inquire and be aware of what is happening via various media

1. *Media and publications* – Stay abreast of mainstream media, academic literature, and publications by governments, professional associations, and NGOs on relevant topics. Determine if there are organizations in your area that already compile this information.
2. *Workshops* – Attend and participate in workshops that are focussed on climate change, particularly those aimed at local governments, including sessions provided by municipal associations.

Take a strong leadership position

1. *Adopt a resolution* – Take a position by adopting a resolution acknowledging the reality of climate change and global warming, and supporting action by the federal and provincial governments.
2. *Support action by other organizations.*
3. *Participate in FCM's PCP program.*
4. *Use policy tools* – Update your official community plans and development bylaws.
5. *Inform and engage your community.*
6. *Set goals* – Commit your organization to achieve specific and ambitious reduction goals, such as "carbon neutrality by 2012."

Smaller communities like the Town of Gibsons do not have the capacity to respond adequately to the issues of climate change and global warming on their own. Assistance is required to effectively sort through public information and scientific reports that are becoming more available to the public. As residents become more aware and concerned about climate change, they are in turn looking to their locally elected officials for answers and direction.

It is fortunate that, as awareness of the issues grows, so too has access to resources, like research staff at local universities and government ministries. There is also greater access to funding, for studies like the Climate Change Impacts and Adaptation Program that was announced by National Resources Canada in March 2007.

As a local government, the Town of Gibsons is becoming more confident in its ability to ask the right questions and to provide reasoned answers for the community. However, if we are to be successful in our efforts to protect the community for years to come, and to be a model for others, it is vital that we remain focussed on this issue.

In 2006, Gibsons council passed a resolution that acknowledges the reality of climate change and its potential devastating effects on communities like theirs; in 2007, it signed the BC Climate Action Charter, committing the town operations to achieve carbon neutrality by 2012, and to pursue green community goals. In taking actions like these, and the concrete follow-through they require, Gibsons council has also encouraged other local governments and orders of government to make this issue a priority and to work together to find appropriate solutions. Council stands ready to partner or work with any other government or agency that has similar concerns, and is willing to share experience and knowledge with others.

Chapter Eleven

Collaborating to Address Climate Change in Edmonton

Michael Evans

The City of Edmonton, Alberta has earned international recognition as one of Canada's greenest cities. Successive city councils have been committed to engaging citizens in addressing climate change, and have been confident that good environmental decision making will make good fiscal sense. Citizens have embraced that same philosophy, and their actions demonstrate their commitment.

In 2006, the City of Edmonton launched EcoVision Edmonton® to bring all of the city's environmental programming – including climate change initiatives – under a single umbrella. EcoVision Edmonton coordinates the city's response to the environmental challenges of the 21st century. It consists of:

- ▶ a new environmental policy that sets the strategic direction for all the city's environmental programs and services, developed through extensive stakeholder consultation;
- ▶ Edmonton's new environmental strategic plan, which includes innovative action against climate change;
- ▶ Enviso, Edmonton's pursuit of ISO 14001 certification for its environmental management systems;
- ▶ numerous programs, including energy efficiency incentives, GHG reductions, and resource conservation measures; and
- ▶ a call to action for all Edmontonians to protect the environment.

By the Community, for the Community

When it first approved a GHG emission reduction plan in 1999, Edmonton council recognized that overall sustained emission reductions were possible only if the entire community was engaged. The city therefore coordinated the creation of a CO_2RE (Carbon Dioxide Reduction Edmonton) "team" that included representatives from over 20 local companies, institutions, and government agencies. The team spent 18 months developing emissions strategies that could be widely implemented. The resultant GHG emission reduction plan was, in effect, developed by the community for the community.

In 2004, the city launched a residential GHG reduction program in partnership with major private partners Home Depot and Epcor, the local power and water utility. The launch included a contest to win one of six $1,000 Eco-Options shopping sprees and 200 $50 rebates for ultra-low-flush (ULF) toilets. Compared to an average month, 88 percent more ULF toilets were sold during the month-long promotion. Sales of efficient products such as compact fluorescent light bulbs, programmable thermostats, insulation, and windows also increased substantially. CO_2RE's residential program launch alone was estimated to have achieved annual GHG reductions of over 1,000 tonnes.

The program was widely supported by local media through paid advertising, public service announcements, and a month-long radio contest. The retail partnership enabled the city to maximize its message with point-of-purchase expertise on energy efficiency, while creating a meaningful channel to distribute the nine titles in the CO_2RE Home$avers booklet series. The booklets covered such topics as attic insulation, basement insulation, and caulking and weather stripping.

In 2005, in recognition of the program's impact, CO_2RE was awarded a FCM-CH2M HILL Sustainable Community Award for sustainable community planning.

New partners have since come on board, and CO_2RE has created new rebate programs for home energy retrofits. Over 10,000 Edmonton households have signed up through the CO_2RE website to receive regular bulletins about practical, in-home solutions to reduce GHGs and save money. The city's experience with CO_2RE has demonstrated that programs to address climate change can be enhanced through collaboration and partnerships between the public and private sectors.

Garbage In, Clean Electricity Out

Edmonton's approach to waste management is integrated and sustainable. The city currently diverts close to 60 percent of residential waste from landfill as part of its 30-year waste management strategy. The Edmonton Waste Management Centre of Excellence delivers some of North America's most advanced waste management practices, including recycling;

composting; household hazardous waste collection; and "e-cycling" – the diversion of electronic items from landfill for re-use. Edmonton's highly advanced waste management practices have prompted it to explore related opportunities that make good environmental sense.

Edmonton is currently the only Alberta municipality to generate electricity by collecting and burning landfill gas, which is a byproduct of solid waste decomposition in landfill. That gas, most of which is methane, would otherwise be released into the atmosphere. Experts report methane is 23 times more potent than carbon dioxide in global warming. Edmonton captures enough landfill gas each year to satisfy the electricity demands of approximately 4,600 homes. Since 1992, 1.5 million tonnes of CO_2 equivalent emissions have been captured – with the effective impact of removing 408,030 cars (209,384 SUVs) from the road.

In November 2006, the city announced it will partner with the Alberta Energy Research Institute to build a $90-million gasification facility to convert residual garbage (waste that cannot be recycled or composted) into biofuel. This facility will support research on the conversion of municipal, industrial, agriculture, and forestry wastes into biogas. The conversion of garbage to clean energy uses a thermochemical process to produce synthetic gas that can be used as an alternative fuel for many purposes. The facility will be operating by the end of 2010.

Conserving Natural Areas

One of the first things visitors to Edmonton notice is the North Saskatchewan River Valley, North America's largest continuous stretch of urban parkland. The 7,400 hectare "ribbon of green" runs through the city's heart, and includes natural areas that are home to deer, porcupines, coyotes, beavers, numerous bird species, fish, and important native plants.

Edmonton was the first municipality in Canada to develop a comprehensive natural areas conservation plan, which indirectly contributes to reducing climate change impacts through the preservation of natural areas.

The *Natural Connections Strategic Plan*, approved by council in July 2007, establishes the strategic policy direction for the preservation and protection of Edmonton's natural areas and native biodiversity. *Natural Connections* is focussed on three intersecting and mutually supportive goals: to secure a functioning ecological network; to manage Edmonton's ecological network; and to engage Edmontonians in environmental protection. As with CO_2RE, involving the public in partnerships to support conservation is a key element of *Natural Connections*.

An example of a recent success has been Edmonton's inclusion of constructed wetlands as part of its stormwater management plan. These ponds, which include native vegetation, provide habitat for important native species, and also help to connect the different elements of the city's functional ecological network.

Natural areas currently comprise nine percent of the city's landmass, an unusually high proportion compared to other municipal jurisdictions. Semi-natural areas – manicured parks, golf courses, transportation and utility corridors, sports fields, and school yards – further increase the percentage of vegetated areas.

Complementary Efforts

Three additional municipal programs that are focussed on energy efficiency and emission reductions merit mention.

In 2001, Edmonton partnered with Natural Resources Canada's Office of Energy Efficiency to develop and deliver Fuel Sense, a program to instruct city drivers how to operate their vehicles for maximum fuel efficiency. Follow-up research determined that, although vehicle use increased the next year by approximately seven percent, fuel consumption per kilometre dropped approximately 5.5 percent – a fuel efficiency gain of 1.8 L/100 km, an estimated fuel cost saving of $175,000, and a reduction of 310 tonnes in GHG emissions. Inquiries about Fuel Sense have come from private industry, all orders of government, and even Madison, Wisconsin.

Another successful initiative has been the city's traffic light retrofit program. The program saves more than 6.4 million kWh per year, and reduces associated power consumption by 80 percent. More than 900 traffic signals, bus destination bars, and lane control fixtures were converted to LED fixtures by 2005.

To further demonstrate municipal stewardship, all new city buildings and major renovations (as of 2006) must meet, at minimum, the Leadership in Energy and Environmental Design (LEED) Silver standard. The city's newest police station was built to meet the LEED Gold standard, and demonstrated that attention to such eco-friendly design and construction could be fiscally, as well as environmentally, responsible. The new standard will enable the city to achieve energy efficiencies 30 percent greater than the Model National Energy Code for Buildings. Edmonton is now working with construction companies and architects to ensure the LEED standard is adopted on all new building projects.

International Recognition

Edmonton's actions on climate change have been recognized across Canada and internationally. Edmonton is one of only two – out of more than

140 – Canadian cities to have received a corporate milestone ranking of five out of five from the Partners for Climate Protection. This ranking places Edmonton among the very elite of the world's cities for addressing climate change.

In 2006, Edmonton was one of only two cities worldwide that was invited to the International Symposium to Establish Responding Strategies to Climate Change in Seoul, South Korea, to share its experience about municipal action to address climate change through the CO_2RE program. Edmonton has also been recruited by ICLEI to be one of 15 international cities in a Local Action on Biodiversity project, recognizing the work it has already done and the leadership it is showing. It is the only Canadian city chosen, and one of two in North America. Other cities include Barcelona, Paris, Johannesburg, and Zagreb.

Conclusion

Edmonton's experience demonstrates that improving energy efficiency, reducing GHG emissions, and mitigating the impacts of climate change are not beyond the capability of municipal governments, especially when citizens and other stakeholders are included in the decision-making process. Edmonton provides a clear example of the tremendous change that can be accomplished within a single municipality when ordinary people know their individual effort can make a difference.[26]

Chapter Twelve

A Unique, Multifaceted Approach

Response of the Montreal Public Health Authorities to Climate Change

Norman King, Louis Drouin, Luc Lefebvre and Tom Kosatsky
Direction de santé publique de l'Agence
de la santé et des services sociaux de Montréal (DSP)

This chapter discusses the response of the public health authorities (DSP) in Montreal, Quebec to the problem of climate change and its impact on health. The DSP's public health mandate requires a multifaceted approach with the objectives of identifying the extent of the problem; helping the vulnerable population adapt to extreme heat; and developing new approaches aimed at improving the urban environment.

The response of the Montreal DSP (hereafter called DSP) to climate change was spawned in the early 2000s by major concerns about health effects from extreme heat events. The DSP initiated a three-pronged response that includes research; a comprehensive heat health warning system (HHWS); and efforts to improve the urban environment through more sustainable development. The response was aimed primarily at protecting certain populations, like the elderly and chronically ill, who suffer the worst health effects from extreme heat.

Achieving progress in sustainable development requires the DSP to work in close collaboration with municipal authorities, both on the central and the borough levels, to ensure that the public health implications of both climate change, and the more frequent and more severe heat waves that will occur as a consequence, are considered in municipal decision making and urban planning.

Heat Waves and Health Effects

As was all too clearly illustrated by severe heat events in Chicago in 1995 and Europe in 2003, extreme heat conditions can have dramatic consequences on health. In Chicago, over 500 people died during a five-day heat wave. In Europe, it is estimated that at least 35,000 premature deaths took place in the summer of 2003, when temperatures soared above their usual range.

Montreal's experience thus far has been less extreme, but significant effects have been seen. It is projected that excess deaths during the summer will increase over time in Montreal, rising from an excess of two percent in 2020 to 10 percent in 2080, and that the excess deaths will affect those aged 65 and over to a much greater extent (Doyon et al., 2006).

Arming Ourselves with Knowledge and Understanding

Our research activities guide our interventions. For this reason, research is critical for building the knowledge base that is essential for effective public health policy responses.

As a first step, it was necessary to define the meteorological conditions under which Montreal citizens are at risk for excess mortality. In order to facilitate an efficient public health intervention aimed at preventing excess mortality, meteorological forecast indicators must be simple to use, and have a high 24- and 48-hour predictive capacity.

Analysis showed that the use of maximal and minimal temperatures over a three-day period were the best predictors (Litvak et al., 2005). The actual threshold levels retained for integration into our HHWS were: a three-day average of maximal temperatures equal to or greater than 33°C; and a three-day average of minimal temperatures equal to or greater than 20°C.

Much work is being done to evaluate whether the DSP's health prevention messages are well understood by the population, and being put into action. A first study looked at people with chronic pulmonary and cardiovascular diseases. It found that, while 70 percent of those questioned seek air conditioned spaces during episodes of extreme heat, a small percentage appears to resist all heat protective advice. The DSP is currently evaluating whether the widespread public education campaign reached the appropriate populations and, if so, whether those populations understood and acted on the preventive recommendations.

The DSP has also examined the extent to which long-term care facilities are equipped (from a clinical and environmental perspective) to adequately protect the chronically ill and elderly during heat waves. Although most of these facilities do not have air-conditioned rooms for residents, the majority do have air-conditioned common areas, where residents can be brought for respite during extreme heat events.

Finally, maps of Montreal's urban heat island have been produced (see Figure 12-1). These maps can be combined with various spatial data, such as neighbourhoods with a high concentration of vulnerable persons. The maps are shared with partners to help refine their various strategies that can impact on health. For example, the municipal housing authority has used these maps to identify seniors residences located in heat island sectors, which should receive priority for air-conditioning installation.

Figure 12-1
Example of a Montreal Urban Heat Island Map

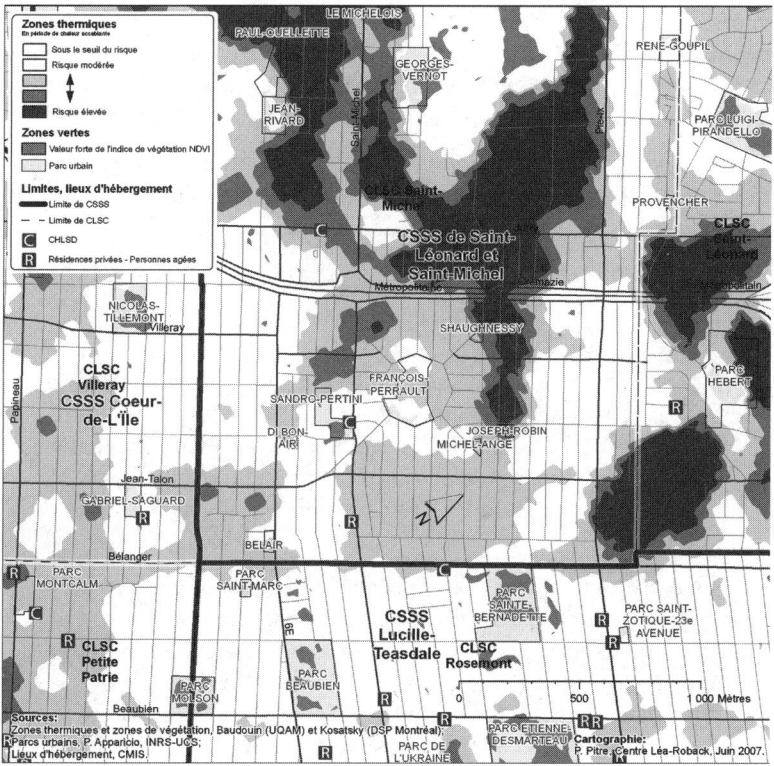

Heat Health Warning System

Montreal's heat health warning system (HHWS) is a comprehensive system that includes a communication plan, surveillance of environmental and health parameters, and emergency responses during extreme heat conditions. Successful implementation of the HWWS requires close collaboration with many partners, including Environment Canada's meteorological service, municipal authorities, and many other organizations that are essential in reaching out to vulnerable populations and implementing various preventive measures

Public education is the first step in implementing HHWS. Starting in mid-May, the DSP delivers targeted communication about health risks from heat waves and about what can be done to prevent adverse health effects. The elderly are specifically targeted with this communication, along with outreach groups such as private residences, medical clinics, pharmacies, municipal structures, and community and humanitarian groups. The DSP used a focus group of elderly citizens to help develop the messaging, in order to ensure it is simple to understand and easy to implement. Media interviews are also carried out during the summer, aimed at the general public as well, urging a community-wide effort to ensure that the elderly and chronically ill are not left alone during heat waves, and that they are able to take preventive measures.

The DSP works closely with Montreal's 12 health and social service centres, supplying information tools that can be shared with their clientele, and supporting them in the elaboration of their local heat plans.

With this groundwork laid, the HHWS can be implemented based on three action levels:

▶ The warning level comes into effect when Environment Canada predicts a temperature of 30°C or more and a Humidex level of 40 or more. By updating the website and maintaining contact with the media, MPH reminds the population and the other groups mentioned above of the preventive message and the importance of outreach towards those who are vulnerable to heat.

▶ An alert level is determined when Environment Canada predicts that the three-day average maximum temperature will reach or exceed 33°C and the three-day average minimum temperature will reach or exceed 20°C. Minimum temperatures that are predicted to stay at 25°C or more for at least two consecutive nights also give rise to the alert level. This level is essentially an internal operation whereby the DSP informs its partners of the potential need to proceed to the intervention level, pending continued monitoring of meteorological conditions, as well as mortality and morbidity data.[27] General information to the public continues to be given through interviews with the media. Daily conference calls with all partners are used to coordinate the planning process prior to the implementation of the intervention plan.

- The intervention plan is initiated when predicted conditions are met or real time surveillance shows an increase in morbidity or mortality. This data is available at 9:00 a.m. for events occurring up until midnight the night before.

The city's boroughs also play a part in the plan by:

- ensuring that public drinking water facilities are fully operative;
- prolonging the opening hours of public swimming pools and wading pools;
- negotiating longer open hours for air-conditioned shopping centres; and
- opening temporary air-conditioned shelters, and organizing transportation to and from these shelters, for those people in greatest need of respite from the heat (the health and social service centres identify this vulnerable population in collaboration with other outreach groups).

Local health centres and long-term care facilities are also mobilized to ensure that their vulnerable populations are able to take adequate preventive measures.

Montreal's HHWS can be effective only if there is close collaboration between the DSP and its partners through all stages, from the planning through to the implementation. This is particularly true for the intervention plan, which clearly requires the mobilization of many resources from different sectors.

Working Towards Sustainable Development

The DSP participated actively with the City of Montreal and other partners in the development of a strategic plan for sustainable development. The plan includes many concrete interventions that could have positive impacts on the urban environment. One example is a pilot project aimed at diminishing the urban heat island effect in a central Montreal neighbourhood through improved urban planning measures (i.e. tree planting, green spaces, green roofs, etc.). The experience gained in this project could then be applied to other neighbourhoods with similar problems.

Over the longer term, the DSP's objectives relate to reducing GHG emissions. This is viewed as the ultimate preventive measure to avoid or minimize health effects from climate change, and also generates many health co-benefits associated with healthier lifestyles and communities. The DSP targeted the transportation sector as a priority for interventions, since it is responsible for 47 percent of GHG emissions in the Montreal area.

The DSP's 2006 annual report, titled "Urban Transportation: a Question of Health," calls for improvements in the public transit and active transportation networks, in order to decrease the population's dependence on the automobile. These calls were reiterated through a public forum held shortly after the release of the report. The DSP also participates in public hearings and consultations on major urban projects, and through them, encourages urban planners to better integrate public transit and active transportation into local developments.

If these interventions are successful, they will simultaneously reduce GHG emissions and generate additional public health benefits, potentially including:

- reductions in air pollution and associated morbidity and mortality;
- reduced traffic accidents; and
- more physical activity, with a corresponding reduction in obesity and associated health problems.

Conclusion

In light of current climate change scenarios, and the already experienced toll of heat waves on human health, it is incumbent on public health authorities to implement measures to prevent excess morbidity and mortality among the vulnerable population during summer heat waves. Also essential is advocacy for changes in the urban environment and its infrastructure that could mitigate heat and reduce GHG emissions to the atmosphere.

The DSP has risen to this challenge by developing a multi-faceted approach in association with a wide range of organizations in the region. In doing so, Montreal is one step closer to a sustainable and healthy city.

Additional Reading and Resources

R.B. Alley, T. Bernsten, N.L. Bindoff, Z. Chen, A. Chidthaisong, and P. Friedlongsten (2007), *Contribution of Working Group I to the Fourth Assessment Report of the Intergovernmental Panel on Climate Change*, Summary for Policymakers.

B. Doyon, D. Bélanger, and P. Gosselin (2006), *Effets du climat sur la mortalité au Québec méridonal de 1981 à 1999 et simulations pour des scénarios climatiques futurs*, Institut national de santé publique du Québec.

D.E. Hémon, J. Jougla, F. Clavel, S. Bellec, F. Laurent, and G. Pavillon (2003), Surmortalité liée à la canicule d'août 2003 en France, *Bulletin épidémiologique hebdomadaire*, 45-46, 221-226.

T. Kosatsky, N. King, and B. Henry (2005). *How Toronto and Montreal (Canada) Respond to Heat*, 167-171; W. Kirch, B. Menne, and R. Bertollimi (2005), *Extreme Weather Events and Public Health Responses*, Springer.

T. Kosatsky, J. Dufresne, L. Richard, A. Renouf, N. Gianetti, J. Bourbeau, M. Julien, J. Braidy, and C. Sauvé, *Heat awareness and response among persons affected by chronic cardiac and pulmonary disease* (submitted for publication).

F. L'Heureux, I. Fortier, A. Smargiassi, N. King, T. Kosatsky (2005), *Profil des mesures d'adaptation à la chaleur et des températures observées en période estivale dans les centres hospitaliers de soins de longue durée, DSP de Montréal*.

É. Litvak, I. Fortier, M. Guoillou, A. Jehanno, and T. Kostasky (2005), *Programme de vigie et de prévention des effets de chaleur accablante à Montréal, Direction de santé publique de Montréal*.

Morbidity and Mortality Weekly Report, Heat-Related Illnesses and Deaths – United States, 1994-1995, 44 (25), 465-468, 1995.

Morbidity and Mortality Weekly Report, Heat-Related Mortality – Chicago, July 1995, 44 (31), 577-580, 1995.

S. Whitman, G. Good, E.R. Donoghue, N. Benbow, W. Shou, and S. Mou (1997), "Mortality in Chicago Attributed to the July 1995 Heat Wave," *American Journal of. Public Health*, 87, 1515-1518.

Chapter Thirteen

Going Green

Sarah Eaton and Geraldine King

The City of St. John's, Newfoundland is the oldest and most easterly city in North America. With a population of 100,000, and a land area of more than 446 square kilometres, the city is the centre for business, research, education, and government for Newfoundland and Labrador. While St. John's is a thriving, modern city with world-class facilities and services, it still has its old world traditions and charm, with a story and a history lesson around almost every corner.

In 2000, St. John's council approved a resolution to enroll in and work through FCM's Partners for Climate Protection (PCP) program. Council was concerned about the potential effects of climate change. At the same time, though, it was excited by the prospect of being the first municipality in Atlantic Canada to complete the PCP program. As council learned more and more about the potential effects of climate change, it acknowledged the need to take action at the municipal level and, as the capital city, to show leadership on this important issue.

That's when the real work of going green got underway.

Getting the Ball Rolling

In 2001, city staff started on Milestone 1 of the PCP process (for an overview of the PCP Framework, see Chapter 14). This would require creating a baseline GHG inventory (based on 1990 base year), and a forecast of 2010 emissions for both the City of St. John's and for the broader community. Developing the GHG inventory was a much more onerous task than originally anticipated. There was insufficient data for a 1990 baseline, so the city instead adopted a 1994 baseline.

The city was fortunate that there was an environmental technician working in the engineering department who was able to dedicate most of his time to the program. Without a dedicated staff person, the program may not have been completed in a timely manner. Small municipalities with limited resources and limited staff may find it a challenge to undertake climate change programs of this type. Having FCM or other agencies provide financial and/or technical assistance is very valuable.

Due to the perseverance and determination of the city's environmental technician, Milestones 1 and 2 (setting a GHG emission reduction target) were achieved in late 2002. The city committed to reducing corporate GHG emissions by 20 percent and community emissions by six percent (from 1994 levels).

The importance of taking climate change seriously was certainly impressed upon St. John's council and the residents in the city's east end in 2002. The city experienced a one in 500-year rainfall event. The associated flooding from the event also helped reaffirm that council's decision to show leadership on addressing climate change was the right one. While people had often joked about global warming – "bring it on, we can hardly wait," or "we'll finally have a warm summer," the changes in weather patterns and increased flooding were a wake-up call that climate change could have serious repercussions for the city and its residents, and indeed, for the whole province.

Getting Our Own House in Order

While the baseline data indicates that corporate emissions are only about two percent of total emissions (community emissions are approximately 98 percent of the total), city officials agreed that, by getting its own house in order, the city would demonstrate to the entire community, including to businesses, industries, institutions, and residents, that reducing GHG emissions could be accomplished.

The city's vehicle fleet is its single largest source of GHG emissions, accounting for some 40 percent of total corporate emissions. Energy use in municipal buildings is the next largest source of emissions. The city therefore undertook several successful in-house initiatives in these areas. More recently, the city started to examine advanced waste management practices.

Municipal building retrofits

In 1995 and 2001, the city retrofitted all municipal buildings. Annual savings are approximately $625,000. By 2007, the total reduction of CO_2 since the project began in 1995 is estimated at 11,000 tonnes. There are average savings of 1,000 tonnes of CO_2 each year, approximately one-third of the total corporate building emissions. With continued monitoring and energy analysis, the city will further improve energy reductions. In 2001, the city received an FCM-CH2M HILL Sustainable Community Award, recognizing the energy savings and environmental benefits as a result of the retrofits.

Anti-idling policy for the municipal fleet

The city's recent anti-idling campaign was deemed a success, setting strict time regulations for vehicle warm-up and idling periods, as well as guidelines for vehicle maintenance that increase efficiency. The city is following on the campaign with a mandatory anti-idling policy for municipal vehicles. A further five percent reduction in vehicle emissions (11,453 tonnes) is projected, all the while saving on fuel and engine wear as well.

LED traffic light replacement

The city's engineering department is in the process of equipping all traffic signals with more efficient LED traffic lights. The total cost is approximately $600,000, with a 3.8 year payback from energy savings. The conversion will result in an annual net reduction of 178 tonnes of CO_2.

Methane capture and green electricity production

The city is currently developing a host of improved waste management practices to achieve multiple environmental benefits, including major GHG emission reductions. The Robin Hood Bay Landfill was recently designated a regional landfill site by the provincial government. This was an important first step toward methane flaring that could reduce GHG emissions by up to 130,000 tonnes, depending on the life span of the landfill. A composting facility, materials recovery facility, household hazardous waste depot, and the methane flaring initiative are all in the planning or design stages. All are due to be online by 2009. A methane-capture power plant, capable of powering 3,000 homes for 10 years, is also being considered.

One of the largest and newest programs – a curbside waste diversion program set to be launched in 2008/2009 – has the potential to significantly reduce the city's GHG emissions from waste. Garbage will be separated into:

▶ fibre products (currently 37 percent of the waste);

▶ wet garbage (currently 30 percent of the waste);

▶ plastics/beverage containers (to be recycled, depending on available markets); and

▶ "real" garbage (residual materials that will end up at the landfill).

The wet garbage will be diverted to a new compost facility at the landfill operation. The fibre material will be recycled; it will be sorted and baled at the new materials recovery facility, which will also be constructed at the landfill operation. By diverting 50 percent of its waste from landfill (the current provincial guideline diversion rate), the city could potentially see a reduction in waste emissions of 50 percent. As the program continues and expands, this rate will also continue to grow.

Tackling Emissions in the Community

The community GHG emission inventory includes residential, institutional, commercial, industrial, transportation, and solid waste sectors. Collectively, these sectors account for some 98 percent of St. John's total emissions, far exceeding the emissions from the city itself. As in most Canadian cities, the largest contributors to GHG emissions in the community are from transportation and residential energy use.

The City of St. John's was again fortunate to be able to hire a climate change intern, in partnership with the Conservation Corps of Newfoundland and Labrador. Having a person on staff that was able to work on the city's climate change program was a definite move in the right direction. Their first task was to help prepare a climate change action plan aimed at achieving the six percent reduction in community emissions to which council was committed. Upon council's approval of the Climate Change Action Plan in 2006, Milestone 3 of the PCP process (developing a local action plan) was reached.

Through the years, St. John's has developed partnerships with non-government organizations and community groups to keep the city safe, clean and environmentally responsible. Continuing to work with community groups to help reduce the amount of GHG entering the atmosphere is an important component of the action plan.

And ... Action!

While past projects have often provided information and built awareness, the emphasis has shifted to more action-oriented projects. The action plan, which is available online,[28] contains detailed action items to reduce GHG emissions in the community. Notable initiatives under the plan that are already underway include the following:

STEER

In 2006, the city partnered with the Taxi Industry of St. John's, Natural Resources Canada (NRCan), the provincial Department of Environment and Conservation, and the Conservation Corps of Newfoundland and Labrador to create STEER – Smart Taxis Encouraging Environmental Respect. STEER is an educational program to help taxi drivers learn about environmentally friendly driving techniques, and vehicle maintenance activities that help reduce both GHG emissions and costs.

The project was the first in Canada to engage the taxi industry on voluntary action to address climate change. Partnerships were the key to the project's success; everyone involved benefited from the mutual respect, trust and support that developed as a result of the project.

The project officially ended in 2006. However, the city was approached by NRCan to develop a multimedia resource and education kit to support implementation of the STEER program in other communities.

RideShare

The city partnered with a student-run volunteer group from Memorial University on its "RideShare" program. RideShare helps facilitate carpool arrangements between people coming to the university from the same area or neighbourhood each day. RideShare is for people who want to travel in a more economical way, respect the environment, save energy and widen their contacts and friends. A website < www.mun.ca/rideshare> was created to highlight the benefits of sharing a ride to work or class.

Other city initiatives will include the establishment of a number of stakeholder advisory committees on transportation, energy, adaptation to climate change effects, as well as a number of public open houses on climate change.

Conclusion

The City of St. John's is committed to protecting its environment and moving towards sustainability. Leading by example, the city is showing other municipalities in Newfoundland and Labrador that climate change demands attention, and that the tools and strategies required are available.

Climate change does not come disguised as longer, warmer, or more pleasant summers. The impacts are often severe and unexpected. This is a lesson St. John's learned the hard way. The good news is that this story is being told, so hopefully other communities won't suffer the same kinds of surprises as a result.

Chapter Fourteen

Planning for Change in Richmond Hill

Andrew Cowan

The Town of Richmond Hill, Ontario is part of a nation-wide movement to make Canada's municipalities healthier, greener, and more sustainable. As a member of FCM's Partners for Climate Protection (PCP) program, Richmond Hill adopted its local action plan in 2004. The plan provides a step-by-step guide to reducing the municipality's impact on the environment by making more efficient use of all of its resources.

Partners for Climate Protection

PCP is a network of more than 150 Canadian municipal governments that have committed to reducing GHG emissions and acting on climate change. The program works in partnership with Cities for Climate Protection, a network of more then 800 communities worldwide making the same efforts, sponsored by the International Council for Local Environmental Initiatives (ICLEI).

Members follow a five milestone framework that includes:

1. Creating a greenhouse gas emissions inventory and forecast;
2. Setting an emissions reductions plan;
3. Developing a local action plan;
4. Implementing the local action plan or a set of activities; and
5. Monitoring progress and reporting results.

Members are supported each step of the way with a variety of tools and resources, such as inventory protocols and spreadsheets, workshops, templates, planning guides, and networking opportunities, all of which are available from the FCM Centre for Sustainable Community Development. Many members have also taken advantage of FCM's Green Municipal Fund (GMF) to finance their progress through some milestones.

By implementing its local action plan, Richmond Hill has now reached Milestone 4, and has begun work on Milestone 5, monitoring the plan's progress.

Taking that First Step

Richmond Hill adopted its first clean air plan in 1998, one of the first Ontario municipalities to do so. After winning a 2000 FCM-CH2M HILL Sustainable Community Award for the plan, the town made a decision to join PCP.

"At the time, everyone was trying to figure out how to respond to climate change and meet the Kyoto Protocol," recalls Dan Olding, Richmond Hill's manager of environmental programs. "PCP gave us a way to break down the obstacles and demystify the process."

Many municipalities in southern Ontario have already experienced climate change impacts, and Richmond Hill is no exception. The town endures an increasing number of smog and heat alert days each year. Flooding from severe storms or spring runoff is also a real danger for Richmond Hill, and the town has responded with a strong stormwater management program. The situation also prompted Environment Canada to partner with the town on a study to evaluate the effectiveness of management techniques in reducing pollution from snow disposal site runoff.

But, it is Richmond Hill's burgeoning population that is, by far, its biggest challenge. In only a decade, the town's population has increased by 70 percent, resulting in more housing developments, more roads, and more services that the town must deliver.

You Can't Manage What You Don't Measure

In 2003, Richmond Hill received funding from GMF to develop energy use and emissions inventories and prepare its local action plan. Creating a GHG emissions inventory and forecast is the first PCP milestone.

ICLEI helped town staff compile years' of energy usage information from buildings, streetlights, fleet vehicles, and water and sewage plants. Olding says that "having the inventory is the most important part of the process, because you can't manage what you don't measure."

With the information in hand, the town determined its energy footprint, and used 2000 as the baseline year upon which to set its targets. Richmond Hill adopted the suggested targets of reducing emissions in municipal operations by 20 percent by 2009, and by six percent throughout the community as a whole by 2010.

Developing an Achievable Plan

The second and third PCP milestones are to set an emissions reductions target, and to develop a local action plan.

It took about two years to develop the town's local action plan, but Olding believes that the time spent was worthwhile. "There's no point bringing a plan to council that isn't achievable, so we stuck to it until we had one that was," he said.

Staff discovered that the greatest opportunities were found in the way the town did business every day. "We use funds that are already dedicated and just work smarter," said Olding. "Most of the plan's recommendations to meet our targets are through regular business practices that don't cost any more money, so council is thrilled."

The plan is fully integrated across all municipal departments, allowing each division the freedom to come up with its own ideas. "In some ways, it's me trying to keep up with them," laughed Olding. "I was recently at a presentation of a new theatre complex that we're building and found that the asset management team is going to be using geothermal as part of the heating system."

The plan is also embedded into the town's ISO 14001 environmental management system. "We're audited on meeting the plan's targets, so if we don't progress, our ISO certification could be challenged."

Sexy Isn't Always Better

On the municipal side, the town's greatest consumers of energy – and therefore the greatest producers of emissions – are its buildings and street lights, representing 92 percent of the total corporate emissions. "The lowest hanging fruit is sitting in our buildings, and we found that the best actions to take were not always the sexiest ones," Olding said.

One of those not-so-sexy initiatives was installing computerized building automation programs to control all heating and cooling systems from a centralized location. "Before, if a building was too hot, it could be up to three days to get someone to look into it. Now, we can track the systems and adjust them very quickly. That's reaping enormous benefits."

Facility improvements, such as lighting upgrades, installing air exchangers and block heaters, and adopting a green purchasing policy, are now par for the course, either as complete retrofits or as part of the town's regular repair and replacement schedule.

The results are impressive. Despite increasing the number and size of municipal buildings to deal with its growing population, the town's buildings have become more efficient over the years, and total energy use and emissions have remained flat. Between 1990 and 2000, retrofits produced savings of some $400,000 annually in avoided energy costs. Measures identified for the future could save an additional $800,000 annually, about 25 percent of the town's total energy budget for its buildings.

Getting Around Town

Emissions from transportation make up about one-quarter of all of Canada's emissions, but are a relatively small proportion of Richmond Hill's corporate emissions. That didn't stop the town from forging ahead with several transportation-related projects, including the use of biodiesel in all of the town vehicles that run on diesel fuel. Olding notes that, even though it costs more, the benefits of lower emissions moves the town closer to its targets. The municipality also boasts a small fleet of hybrid vehicles for staff use, and plans to purchase more hybrids as older vehicles are retired.

Partnering with York Region and Markham, Richmond Hill joined Smart Commute, a transportation management association that represents nine municipalities in the Greater Toronto Area and Hamilton. Smart Commute helps local employers explore different commuter options, such as public transit, carpooling, and telework. About 10 percent of municipal employees now carpool, and the town is working to expand its telework program. So far, the partnership has resulted in 19.5 tonnes of GHG reductions and 85,950 fewer kilometres in vehicle trips.

One of the more creative ways the town has found to reduce the number of cars on the road was through the launch of its Lunch Express bus. "Traffic is horrendous over the lunch hour and restaurant owners were complaining that there wasn't enough parking," Olding explained. "We came up with the idea of a free Lunch Express bus that has a set route past major businesses and restaurants." Restaurants supported the program from the beginning, offering discounts to diners who use the bus, and the town plans to expand the route.

Staying on Track

Implementing a local action plan, monitoring progress, and reporting results are the fourth and fifth PCP milestones. Olding says that the PCP protocol is key in evaluating the town's progress. "Using the protocol and creating the inventories help us measure our progress towards our targets," he said. To ensure that they stay on course, the town updates its municipal inventory annually and its community inventory every two years.

Partnerships Matter

Richmond Hill hasn't done it alone. Partnerships with York Region, Markham and the Greater Toronto Area (GTA) have all played a role in the town meeting its objectives. All of the municipalities in the GTA are moving forward and it's a coordinated effort, notes Olding. "We are always looking at what other municipalities are doing and what has been successful," he says. He points to Markham's anti-idling initiative, which Richmond Hill emulated when developing its own policies, and the energy management software the town uses, which was developed by York Region.

Community Engagement

Olding is confident that Richmond Hill will meet its corporate target by 2009. He is less sure when it comes to meeting their community target of lowering emissions by six percent by 2010.

That's because the municipal share of energy use and emissions accounts for very little of the overall total. All of Richmond Hill's municipal facilities account for less than two percent of the town's total energy consumption, leaving the lion's share in the hands of the community. "We've made good progress, but we need to push it further," he says. Public displays and open houses only go so far, he notes, so the town is considering other options. One example is to display information about energy efficiency and retrofits at places where people make their decisions, such as when they are getting a building permit.

"We don't have the resources to offer incentives to businesses or individuals, but we can make them aware of other federal, provincial, municipal or utility programs that exist," says Olding.

Powering the Future

As Richmond Hill continues to grow, it will need a steady supply of ideas to keep energy use and emissions in check, while still providing a high quality of life for the town's residents. Pursuing renewable energy options ranks high on that list.

Wind power, for instance, will likely play a big role in new electricity generation. A year-long monitoring study showed that three five-megawatt turbines would provide a modest return on investment and power 2,500 homes. Staff are also looking into using solar energy for town hall, and are exploring the use of district energy for a new campus-type development in Richmond Hills's old town core.

"Every year we have more employees, more buildings, more residents, so we have to provide more services," said Olding. "But, we stuck to our main sources of funding and thought creatively. It all flowed from there."

Conclusion

The Partners for Climate Protection program is intended to catalyze community action on climate change and support sharing the wealth of knowledge that is already available and growing every day. By providing a systematic, tried-and-tested process for building a local climate change response, PCP can assist in building the local case for action and assist development of corporate and community emission reduction plans.

Cities do not need to act in isolation on climate change – opportunities abound for collaboration and sharing with those next door, across Canada, and around the world. There are many communities, such as Richmond Hill, on the leading edge of innovation. Each has much to share about their experiences. With a focus on initiatives with high transferability, turning to the PCP program is a good first step and a continuing resource if your community is looking to take action on climate change.

Additional Reading and Resources

Partners for Climate Protection: Richmond Hill's local action plan can be downloaded from the PCP website at <www.sustainablecommunities.fcm.ca>.

FCM-CH2M HILL Sustainable Community Awards, 2000 to 2006 Best Practices Guide can be downloaded from <www.sustainablecommunities.fcm.ca/Programs/FCM-CH2M_Awards>.

Richmond Hill's Sustainable Stormwater Management Program is online at <irc.nrc-cnrc.gc.ca/pubs/bsi/2006/cd/fulltext/NewRichmondHilllongCaseStudy2006.pdf>.

Smart Commute information is online at <www.smartcommute.ca>.

InfraGuide: The National Guide to Sustainable Infrastructure is hosted at <www.sustainablecommunities.fcm.ca/infraguide>.

ICLEI: Information about ICLEI and Cities for Climate Protection is online at <www.iclei.org>.

Chapter Fifteen

Using Climate Scenarios at the Municipal Scale

Caroline Larrivée and Guillaume Simonet

Climate change scenarios are best understood at the global scale. At the local or regional scales, the uncertainties about the magnitude, direction and rate of climate change are greater. This uncertainty can make it difficult for local-level decision makers to fully grasp the level of vulnerability to which their communities, infrastructure, and socio-economic activities are exposed. It can also have an impact on their willingness or ability to implement adaptive measures.

However, local actions on climate change cannot wait for perfect science. Adapting to an uncertain future climate is something we must learn to do.

Despite the range of possible outcomes, climate scenarios can be a useful decision support tool for helping to understand and cope with the challenges that climate change creates. This chapter explores how climate scenarios can support adaptation policy development at local community levels by indicating the various types of impacts linked to climate change that are likely to occur, helping to make the anticipated changes more tangible.

Rethinking Current Ways of Designing and Developing

Changes in long-term temperature averages, climate variability, and in the frequency, duration or intensity of extreme events can have direct and indirect consequences on the natural and built environment, on populations, and on socio-economic activities. Figure 15-1 presents examples of such impacts for municipalities.

Figure 15-1
Direct and indirect impacts of climate change at local scale

Systems	Climate Event	Direct Impacts	Indirect Impacts
Infrastructure	Extreme precipitation events	Stormwater drainage systems are incapable of collecting excess run-off	Houses are flooded, mudslides destabilize infrastructure
Communities/ populations	Heat spells in summer	Discomfort for vulnerable populations	Decline in welfare of population with psychological, medical, and social consequences
Socio-economic activities	Reduced snowfalls in winter	Reduced patronage of ski resorts	Increased pressure and competitiveness between ski stations
	Longer drought periods	Agricultural losses for individual farmers	Reduced vitality for local and regional economies dependent on these activities
Environment	Increased temperatures in summer	Water temperatures in lakes and rivers increase	Quality of source water decreases and requires optimized treatment
Cumulative and domino effects	Ice storms	Transmission lines collapse and cause power failures	Water treatment and distribution systems dependant on power supply are affected

At the same time, our communities are affected by and respond to a multitude of non-climate related factors that can make them more or less vulnerable to new climate conditions.

Most land use planning and infrastructure design criteria factor in climate considerations to ensure structures are built and located to be safe and resistant to anticipated weather conditions. Such criteria take into account the level of risk that is judged acceptable with respect to construction costs.

This approach assumes that past climate conditions in a given area are representative of future climate conditions, and that this holds true for the life span of the infrastructure. Hazardous areas, such as flood plains and areas prone to landslides, are assumed not to vary with time or with extreme weather events.

But, the climate is changing. Different regions of Canada have already registered statistically significant temperature increases in the past 50 to 100 years. Many communities have experienced the effects of a warmer climate through melting permafrost in the north, accelerated coastal erosion in maritime areas, or more frequent summer heat spells and extreme precipitation events throughout the country.

This forces us to rethink our approach to planning and design. The use of climate statistics based on historical observations no longer seems a suitable approach.

Given these uncertainties about climate changes and their implications at the local level, how can we account for climate change in our planning and design activities? How can we use this to design and implement strategies to reduce our vulnerability?

One approach is to use climate scenarios. Climate scenarios can be used in combination with other decision-support tools, such as infrastructure vulnerability assessments, cost/benefit analyses, hazardous area maps, or implementation of no-regret measures (eg. water smart programs).

What are Climate Scenarios?

Climate scenarios are representations of plausible future climates, which are constructed to help investigate the potential consequences of human-induced climate change. These scenarios can provide information on what kinds of changes are likely to occur in terms of temperature, precipitation, winds, or soil moisture, for example. Scenarios are based on statistical averages and probabilities calculated from weather stations covering a given area.

Climate scenarios use mathematical models to reproduce weather systems and conditions to simulate the climate. Both global circulation models and regional climate models are used to provide insight into what the future climate might be like within a 100-year time frame.[29]

Using Climate Scenarios at Municipal Scale

At the municipal level, climate scenarios provide an indication of what climate conditions are likely to be encountered, presenting impacts as more tangible concerns. They help create awareness about climate change risks and associated impacts, and highlight the importance of implementing mitigation and adaptation actions. Indeed, scenarios can be used to build strong justifications for implementing strategies that reduce vulnerability. For example, scenarios can identify new hazardous areas (areas of coastal erosion, areas prone to flooding, areas exposed to heat island effects), and inform policies to regulate or restrict development in those areas.

Climate scenarios can also be used to demonstrate the need to revise methods by which climate variables are currently factored into construction design and community development.

Climate scenarios provide information on physical vulnerabilities to climate change. Other data and information, such as socioeconomic and demographic data and local indicators, can provide insight into adaptive capacity and social vulnerabilities. All of these factors are important for climate change decision making, since they jointly determine a community's overall vulnerability – and hence its need to adapt.

Climate scenarios are therefore best used in combination with other decision-support tools. Many of these tools are commonly used in local planning and design. For example:

- ▶ IDF-curves, which are used to establish the dimensions of stormwater management networks, can be revised based on climate scenarios to see how the frequency of extreme events might increase. Municipal engineers can use this information to better plan drainage infrastructure.

- ▶ In more northerly locations, climate scenarios can be used with soil maps and projections of soil temperature profiles to help identify areas where melting permafrost degradation may cause damage to buildings or infrastructure.

- ▶ Scenarios can be used in conjunction with zoning maps to identify and illustrate anticipated rates of erosion in coastal areas, and to plan or adapt developments accordingly (eg. by prohibiting construction in certain more vulnerable areas, relocating existing infrastructure that is especially vulnerable, or reloading beaches to slow down coastal erosion).

- ▶ Combined with land use maps to identify areas at risk of heat island effects, and combined with demographic data, scenarios help target prioritized solutions to vulnerable populations.
- ▶ A community's expected climate conditions can be compared against a current analogous climate in a similar community located further south. The community can then begin implementing comparable strategies, such as construction techniques, building materials, urban tree species, etc., in anticipation of future conditions.

All of these decision-support tools can be used to help identify potential adaptation strategies. Strategies can then be assessed using cost-benefit analysis or similar methods to determine the best option.

Figure 15-2
Adaptive Capacity of a System

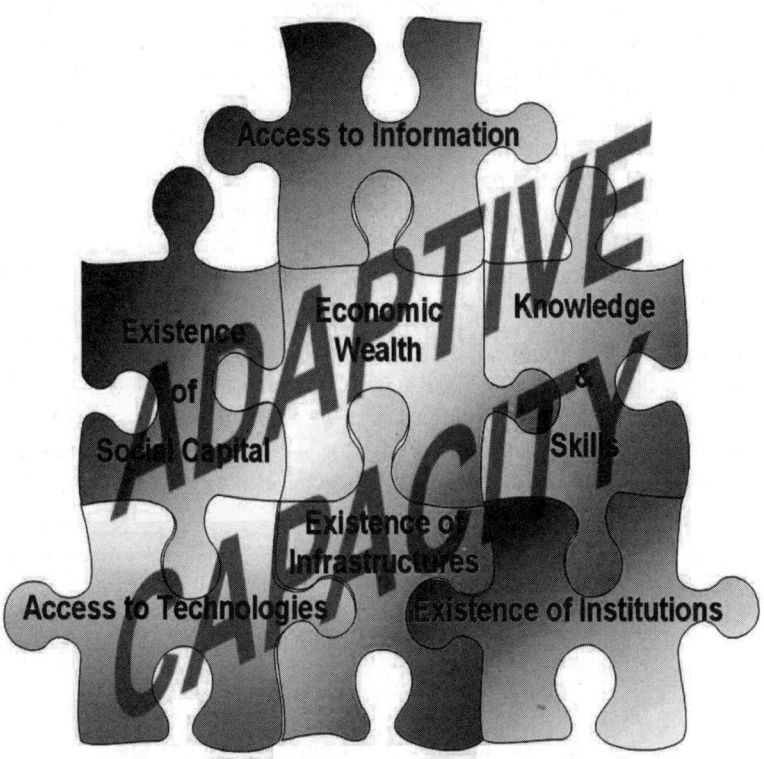

Adaptive capacity and some of its essential components, from Ouranos, adapted from Fussel and Klein (2005).

Limitations of Scenarios

It is important to keep in mind that scenarios are neither a prediction nor a forecast of the future, but a plausible indication of what the future could be like over decades, given a specific set of assumptions.[30] Due to inherent natural variability of the climate and the complexity of the interactions between various components of the natural environment, there are many uncertainties associated with climate scenarios.[31] Indeed, systems such as El Niño or other cycles that can affect weather conditions over several years make it difficult to assess whether specific weather events are related to climate change. Also, the adaptive capacity of a system will depend on the various components that make up the system. As Figure 15-2 illustrates, characteristics such as a thriving economy, strong political will, and access to information and technology increases the capacity to adapt to changing conditions.

Conclusion

Current approaches to land use planning and infrastructure design use climate criteria drawn from historical weather data, based on the assumption that past weather conditions for a given area are representative of future conditions. Given that the climate is changing, this assumption must be revised.

Climate scenarios assist local and regional decision makers to identify potential future climate impacts for their specific communities, and guide the implementation of adaptation strategies. Despite the uncertainties surrounding future climate conditions that require caution in using climate scenarios, such scenarios remain important decision-support tools that can guide and improve planning decisions and construction techniques.

As one component within a multitude of decision-support tools, climate scenarios can provide municipalities with crucial information necessary to implement adaptation strategies relevant to their community. Adaptation strategies and actions help reduce the vulnerability of communities, and help them cope with the challenges generated by climate change.

Additional Reading and Resources

There are a variety of ways to access climate scenarios for different climate variables that range from global to more regional or local scales.

The IPCC reports summarize the anticipated climate changes for regions throughout the world. For more regional coverage, there are a number of resources in Canada, depending on your specific location. These organizations can sometimes provide scenarios for specific locations upon request. Such organizations include:

- Environment Canada through the Canadian Centre for Climate Modelling and Analysis <www.cccma.ec.gc.ca> and Climate Change Scenarios Network <www.ccsn.ca>.

- Ouranos <www.ouranos.ca> is a research consortium whose mission is to advance the understanding of regional climate change and of its environmental, social and economic impacts.

- Pacific Climate Impacts Consortium <www.pacificclimate.org>.

- Canadian Climate Impact Scenarios <www.cics.uvic.ca/scenarios>.

- Prairie Adaptation Research Collaborative <www.parc.ca>.

References

S. Dessai and M. Hulme (2004), "Does climate adaptation policy need probabilities?" *Climate Policy,* 4: 107-128.

H.M. Fussel and R.J.T. Klein (2005), "Climate Change Vulnerability Assessments: An Evolution of Conceptual Thinking," *Climatic Change* 75: 301-329.

IPCC (2000), *Summary for Policymakers Emissions Scenarios*, A Special Report of Working Group III of the Intergovernmental Panel on Climate Change.

Chapter Sixteen

Green Municipal Fund

Facilitating Local Learning and Action

Elisabeth Arnold

The Green Municipal Fund (GMF) is a key strategic resource to help municipal governments achieve their sustainability goals. The Government of Canada endowed the Federation of Canadian Municipalities (FCM) with $550 million to establish the fund and provide a long-term, sustainable source of financing for municipal governments and their partners.

Through the fund, FCM provides low-interest loans and grants, builds capacity, and shares knowledge to support municipal governments and their partners in developing communities that are more environmentally, socially, and economically sustainable.

FCM maximizes the impact of GMF by leveraging each investment to build new knowledge and skills on sustainable infrastructure. FCM can then transfer these lessons to other municipalities and foster change. By providing municipal governments with the funding and knowledge they need to meet their goals, FCM is leading sustainable municipal development in Canada.

GMF Overview

The GMF was established in 2000, when the Government of Canada endowed FCM with $125 million. This was doubled in 2001-2002 to $250 million, and increased again in 2005 by $300 million.

Funding is allocated to sustainable development projects in six sectors of municipal activity: brownfields, energy, planning, transportation, waste, and water. Grants for feasibility studies, field tests, and sustainable community plans are provided throughout the year, while grants and loans for capital projects are awarded through a competitive request for proposals process. The very best capital projects – with high environmental, social, and economic benefits – are eligible for a combination of grants and loans worth up to 80 percent of project costs.

By April 2007, GMF had funded more than 600 projects and studies. That $320-million investment in municipal sustainability has leveraged more than $1.7 billion in additional investment from other sources. The initiatives will deliver significant environmental benefits, including reduced GHG emissions of over 1,200 tonnes of nitrogen and sulphur oxides, and 1.2 million tonnes of carbon dioxide equivalents.

To ensure the greatest possible impact, GMF is strategically invested in projects with the most significant potential environmental benefit, and in projects that can be replicated in other municipalities.

While the GMF is not just a climate change fund, there are climate change impacts from projects undertaken in most of the sectors supported. The most significant GHG reductions have been within the energy sector, which accounts for more than one-third of financing offers in the fund's first six years. GMF has supported small and large projects, ranging from a small feasibility study for the Nelson and District Aquatic Centre Solar Retrofit (total project value $8,000), to the implementation of a deep lake water cooling system (total project value $176 million) in Toronto. Sustainable transportation projects have also been developed with attention to GHG reductions – including studies on the inclusion of hybrid vehicles in municipal fleets, ride-matching services, transportation demand management, and community-wide transit passes. In addition, a number of municipalities have received GMF support to develop comprehensive community climate change plans in order to meet their commitments as members of Partners for Climate Protection (PCP).

From Financing to Knowledge

There are many municipalities at the leading edge of sustainable development, with projects that promise significant environmental benefits. A great many more are somewhere closer to the middle of the pack.

GMF focuses on funding the projects and studies at the leading edge of the curve. These projects are not necessarily those using the newest technologies. Rather, they are innovative in applying technologies, processes, or governance models in a municipal context, and serve as preeminent examples that other municipal governments can learn from and follow.

FCM collects the lessons learned from each of those leading studies and projects and transfers that knowledge and experience back to the middle of the pack, so that other municipalities can also advance. As important as disseminating the lessons learned is, however, putting communities in touch with others who have on-the-ground experience is critical. Relationships help to bridge the gap between knowledge and action on sustainable development.

Knowledge Support Services

GMF's capacity building program was launched in 2006. It is helping the fund to make the transition from being solely a financial tool to support sustainable change in individual municipalities, to becoming a means to leverage change in all Canadian municipalities.

The capacity building program informs, inspires, and supports municipal governments to implement sustainable community development projects and practices. It includes campaigns in each of the six sectors of municipal activity funded by GMF. Recent work with the International Institute for Sustainable Development has helped define effective tools and approaches for sharing knowledge. That research will help create the framework on which the six campaigns are based.

With this new addition, GMF has become a municipal leadership program, pushing forward the boundaries of sustainable community development by engaging municipal leaders to share their knowledge gained through GMF projects with other communities, and helping them build their capacity to achieve their own sustainability goals.

In addition to the six campaigns, the capacity building program includes the PCP program, the FCM-CH2M HILL Sustainable Community Awards, the annual Community Energy Planning Missions, and the Sustainable Communities Conference. Each of these programs is designed to further extend FCM's ability to help municipalities access each other's considerable expertise.

Full Cycle Municipal Engagement

By adding knowledge and information to its list of resources, GMF is able to support communities at many different stages of their sustainable development work. The Town of Okotoks, Alberta offers a good example.

Okotoks became one of the first municipalities in the world to establish growth targets linked to infrastructure development and environmental carrying capacity when it adopted a Municipal Development Plan in September 1998. Not content to rest on its achievements, the town began to look for ways to reduce GHG emissions through the creation of an alternative energy program.

In 2002, Okotoks Municipal Manager Rick Quail participated in the FCM's Community Energy Planning Mission to Denmark. The following year, Quail also participated in the mission to the Netherlands. According to Quail, "These energy missions provided valuable and practical insights into the development of district energy systems utilizing renewable energy sources."

In 2003, Okotoks applied for and received GMF funding for a study on the "Feasibility of Large Scale District Solar Heating Systems Utilizing Seasonal Storage." The following year, Okotoks received over $3 million in GMF grants and loans to pilot a 50-home seasonal solar district heating demonstration project, based on a European concept that Natural Resources Canada had been looking for an opportunity to test. The intent was to build energy efficient housing that would get approximately 90 percent of its required heat and hot water from a district thermal solar heat source.

Okotoks received a 2006 FCM-CH2M HILL Sustainable Community Award in the energy category for its solar initiatives. In addition to profiling their project at the awards ceremony, a video of the Okotoks experience and that of the other 2006 award winners was sent to over 1400 Canadian municipalities. In January 2007, Okotoks' solar energy success story was featured on the cover of *Municipal World*. And, in the summer of 2007, participants in FCM's 2007 Community Energy Planning Mission to Alberta visited to further discuss how renewable energy supplies and district energy systems could be implemented across the country.

The capacity building program is fortunate to be able to count on the experiences of innovators and early adopters of climate change programs, like Okotoks. Municipalities that have received GMF support in the past, through knowledge and financing, are now reaching out to other municipalities to share their experiences.

Expanding Community Engagement

In many communities, the context for municipal decision making is increasingly shifting from a command-and-control style of public administration to a more complex, networked style of community decision making and leadership. Municipalities are increasingly demanding not only knowledge about sustainable development innovations and opportunities, but also that such knowledge be delivered in such a way as to enable them to engage their local partners within the community.

Stephen King, Manager of the Sustainable Environment Management Office with the Halifax Regional Municipality (HRM), observes, "Many municipalities now want to move beyond the education and planning stages, to operationalizing and/or implementation. This is why we are working so hard on sustainability governance models and partnerships to help us prepare."

From 2002 to 2005, the HRM received a series of GMF grants to undertake studies on the development of district heating systems and community-wide climate change mitigation and adaptation opportunities. To facilitate these projects, the municipality embarked on building a Sustainable Community Corporate Environmental Management Plan and received strategic sustainability training to help it develop a holistic approach to sustainable development.

With these foundations in place, the municipality embarked on the Climate-SMART (Climate Sustainable Mitigation and Adaptation Risk Toolkit) project, launched in March 2004 to build a local climate change action plan. The action plan sought to be broad-based, encompassing not only HRM's significant corporate assets and activities, but also those assets and activities within the community at large, over which HRM has substantial indirect influence. Dan Sandink shares more details on Climate-Smart in Chapter 6.

Halifax recognized that engaging local organizations in climate change efforts would be critical to achieving real GHG reductions. However, this goal is not an easy one to reach. As King notes, "The biggest gap is with the foundation – getting it established within the community. When we first started talking, you could see people's eyes getting glazed over. It's good to have peer exchanges and knowledge transfer, but it's also important to have a project in your own community that people can feel and touch. We've found that the ground up approach is essential. You also have to take full advantage of sustainability resources and opportunities in your own community."

HRM's Climate-Smart initiative built strong partnerships between all three orders of government, community groups, local business, and the private sector. Together, they have been able to coordinate efforts and apply their shared resources towards a truly integrated, comprehensive approach to climate change. Like Okotoks, HRM shared it sustainability story in the pages of *Municipal World*, appearing in the August 2007 issue.

Conclusion

Strategies for climate change adaptation and mitigation can be a framework that ultimately leads to the development of more sustainable communities. These are the kinds of communities that FCM wants to foster and help build. FCM's capacity building program aims to expand the opportunities for municipal staff and their local partners to learn from peers – like Rick Quail and Stephen King.

Case studies are systematically generated from all GMF-funded initiatives, and national best practices are disseminated through an awards program, publications, workshops, and a website. As GMF increasingly invests in advanced projects with high replicability, its bank of lessons and knowledge will grow, developing FCM as an important centre for municipal sustainable development learning in Canada. The GMF program is thus a key resource for communities at all stages of understanding and action.

Additional Reading and Resources

Green Municipal Fund:
<www.sustainablecommunities.fcm.ca/GMF>.

Halifax – Naturally Green: <www.halifax.ca/environment>.

Sustainable Okotoks: <www.okotoks.ca/sustainable/overview.asp>.

Epilogue

Susan M. Gardner

Even as work was underway in wrapping up this publication, inspirational stories were emerging from across the municipal sector, with municipalities engaging and moving on the climate change issue in ways that many other governments have failed to do.

One of the highlights was the announcement at the September 2007 conference of the Union of British Columbia Municipalities that 62 local governments in the province had signed a Climate Action Charter with the provincial government and the UBCM. Signatory local governments committed themselves to developing strategies and taking actions to become carbon neutral in respect of their operations by 2012; to measure and report on their community's GHG emissions profile; and to create complete, compact, more energy efficient rural and urban communities (for example, by fostering a built environment that supports a reduction in car dependency and energy use, by establishing policies and processes that support fast tracking of green development projects, and by adopting zoning practices that encourage land use patterns that increase density and reduce sprawl).

BC Premier Gordon Campbell used the UBCM convention setting to commit his government to take steps to mandate greenhouse gas reduction targets, and to provide the legal tools that will be necessary in order to reduce greenhouse gas emissions by one-third below current levels by 2020. He also announced that greenhouse gas emissions reduction strategies and targets will be legally required in all official community plans and regional growth strategies, and that municipalities would be empowered to waive development cost charges to help encourage green developments, small unit housing and small lot subdivisions.

If we are going to be successful at combatting the climate change problem, we need more initiatives like the one in BC – we need initiatives that are aggressive, challenging, and courageous. As David Noble and Trevor Dixon Bennett point out in the opening chapter, this problem is a wicked one. The amount of CO_2 emitted by Canadians in the course of their daily living is astonishing, and comes from a multiplicity of sources. It ranges from our housing and work environments, to transportation, to manufacturing, to agriculture ... just about every facet of human activity. Luckily, though, many of the levers to control it are policy-based. This requires the establishment of targets, and the creation of policies and strategies to meet those goals. It requires leadership and action from those entrusted with protecting the health and well-being of Canadians.

Those who work on the front lines of local government, whether elected or appointed, know that climate change isn't happening in a vacuum. It's happening within the context of a whole range of other important local issues. But, the co-benefits of tackling the climate change problem are numerous. As just one example, by encouraging greater use of public transit (and taking the requisite steps to make it available, accessible, etc.), communities not only reduce CO_2 emissions from transportation, but also reduce gridlock, and help to create pedestrian friendly communities that contribute to a healthier population, safer streets, and so on.

The Ontario Professional Planners Institute recently released a paper, *Healthy Communities, Sustainable Communities*, that refers to healthy communities as the "21st century planning challenge," encouraging us to pay attention to a growing body of research on what differentiates a healthy neighbourhood from a less healthy one, and to "apply this knowledge in our work, whether we are planners, health professionals, educators, social service providers, or decision makers." While the OPPI paper is not targeted to address the climate change issue *per se*, the actions it advocates tie in nicely with mitigation and adaptation strategies, providing another very clear example of the co-benefits that can accrue from a strategic, coordinated approach to local issues. Small positive changes in one area can have significant ripple effects across numerous areas. The OPPI paper points to the need for municipalities to breakdown the silos within and across their organizations, to develop multi-disciplinary strategies and solutions. Alex Boston's rendering of this principle in Chapter 4 as "Top Down, Bottom Up, Across and Inside Out" is an apt description of the organizational changes needed to perforate the silos that keep us from finding the best solutions for the planet.

Municipalities are taking up this challenge, and are beginning work within and beyond their organizations to share knowledge, ideas, and strategies. In 2007, climate change was prominently profiled on the agenda at numerous municipal conferences here in Canada, including the annual conferences for UBCM, FCM, the Canadian Institute of Planners, the Association of Municipalities of Ontario, the Manitoba Planning Conference, the Transportation Association of Canada, and the Union of Nova Scotia Municipalities, to name just a few. Internationally, events like the C40 Large Cities Climate Summit in New York City and the Global City Forum in Lyon, France were successful in bringing together municipal leaders from around the globe – decision makers determined to give the issue traction, and to take the concrete actions required to achieve measurable results in their communities.

Canadian communities are stepping up with both local action and collaborative activity as well. As an example, the Clean Air Partnership recently announced the formation of the Alliance for Resilient Cities (ARC), with a goal of "bringing together government officials and scientists from Canada, the United States, and England to discuss the potential catastrophic impacts of global warming on urban areas and the adaptations cities must make to protect their communities." The inaugural symposium for the ARC included presentations from Halifax Regional Municipality, City of Peterborough, Metro Vancouver, NRCan, the UK Climate Impacts Programme, and King County, Washington. This is an example of how knowledge and best practices on climate change can transcend community, provincial and even national boundaries, to provide meaningful lessons across those divides – much as this book aims to do.

Municipal World has been pleased to partner with the Federation of Canadian Municipalities, 2degreesC, and other sponsors in sharing the critical messages in this book – that it was deemed important and necessary is a tribute to the power of local governments to truly move the world, by shaping their communities to effect change. To paraphrase the famous quote of Mahatma Gandhi, Canadian municipalities are taking steps to "be the change" they want to see in the world. To be successful, they will need to move with leaps and bounds, and with ambitious efforts to lead the work necessary to protect and preserve their communities for a sustainable future.

Endnotes

1. To the IPCC, "very likely" means that an outcome or result is more than 90 percent likely.
2. Greenhouse gases "absorb" energy, rather than "trap" it, but the effect is the same – GHGs in the atmosphere cause the atmosphere to warm.
3. *The Globe and Mail*, April 4, 2007, "Forecast for Prairies: drier than a dustbowl."
4. Sir Nicholas Stern (2006), *Stern Review On The Economics Of Climate Change*, HM Treasury, Government of the United Kingdom, London.
5. 450 ppm means around a 50 percent chance of keeping global increases below 2°C. At this level, is it unlikely that increases will exceed 3°C. At 550 ppm, there is around a 50 percent chance of keeping increases below 3°C.
6. Alex Boston. "Climate roadmap must feature local governments," *Municipal World*, September 2007.
7. Unfortunately, it seems that "just as the impacts of climate change are increasingly being experienced, Canadian governments are pulling apart." In GB Doern (2007), *Innovation, science, environment: Canadian policies and performance*, 2007-08, McGill-Queens University Press.
8. For example, "individual knowledge" based on lived experience and reflection, vs. "scientific knowledge" based on scientific inquiry, vs. "traditional knowledge" based on shared experience, traditions and cultural expectations.
9. Kevin Lynch, Clerk of the Privy Council (April 2006), cited in "Improving the quality of life of Canadians through natural resources," Natural Resources Canada, November 2006.
10. This survey, entitled "The Local Politics of Climate Change," was conducted by Devin Causley in support of a graduate thesis at the University of Waterloo. Data was gathered in the spring of 2007 resulting in 118 responses from municipal staff and elected officials across Canada. Questions explored the decision-making process by municipal officials related to climate change.
11. Alex Boston (2006). "Best Process Before Best Practice: Lessons from Leading Cities on Climate Protection." Dissertation highlights. Presentation to the Federation of Canadian Municipalities, Ottawa, October 2006.
12. Michele Betsill (2001), "Mitigating Climate Change in US Cities: opportunities and obstacles." *Local Environment*, Vol. 6, No. 4: 393-406.
13. Harriet Bulkeley and Michele Betsill (2005), "Rethinking Sustainable Cities: Multilevel Governance and Urban Politics of Climate Change," *Environmental Politics*, 14 (1), pp. 42-63.
14. D. Cromwell, D. Edwards (2006). "Chapter 10: Climate Change, the Ultimate Media Betrayal." *Guardians of Power: The Myth of the Liberal Media*. Pluto Press, London.
15. M. Macalister (2004). "Chapter 3: Avenues of Participation in Local Governance." *Governing Ourselves: the Politics of Canadian Communities*. UBCM Press, Vancouver.
16. Federation of Canadian Municipalities (2006). *Enlisting Municipal Governments in a National Approach to Clean Air and Climate Change*. Ottawa.

17. There are more than 130 members of the FCM Partners for Climate Protection program. At the 2006 FCM general assembly, Canadian municipalities declared a commitment to 80 percent reductions by 2050.
18. Harriet Bulkeley and Michele Betsill (2005), *Cities and Climate Change*. Abingdon: Routledge.
19. Loren Lutzenhiser, Kathryn Janda, Rick Kunkle and Christopher Payne (2002), *Understanding the Response of Commercial and Institutional Organizations to the California Energy Crises*, Consultant Report for the California Energy Commission.
20. These equations are variations from the Lutzenhiser et al. study, ibid.
21. Michele Betsill (2001), "Mitigating Climate Change in US Cities: opportunities and obstacles." *Local Environment*, Vol. 6, No. 4: 393-406.
22. The following source provides very good insight into the financial challenges: Ralph Torrie (1998), *Municipalities Issue Table Foundation Paper*, prepared for the Canadian Government's National Climate Change Process.
23. Standards Australia. (2004). *Australian/New Zealand Standard in Risk Management* (AS/NZS 4360:2004).
24. A popular CBC family television series that celebrates its 35th anniversary in 2007.
25. Winner of the 2005 Berkley Springs Water Contest.
26. For more information about any of Edmonton's environmental programs, contact Edmonton's Office of Environment at <env@edmonton.ca>, or visit the city's website at <www.edmonton.ca> and select "Environment" from the menu at top.
27. Hospitalizations or emergency room consultations.
28. <www.stjohns.ca/cityservices/environment/climatechange.jsp>
29. The main difference between global circulation models (GCMs) and regional climate models (RCMs) is the resolution at which each type of model reproduces land and ocean characteristics. The Canadian Regional Climate Model's grid-size is in the order of 45 km horizontally, compared to approximately 350 km for GCMs.
30. IPCC, 2000.
31. Atmosphere, hydrosphere, lithosphere, cryosphere, biosphere.

About the Editors

Susan M. Gardner is Executive Editor of *Municipal World*, where she researches, writes and speaks on sustainability issues and frameworks for the municipal sector. She holds a Masters of Public Administration in Local Government from the University of Western Ontario, and is a member of the Smart Growth Canada Network Advisory Board.

David Noble is the founder and principal of 2degreesC, a consultancy that supports client leadership and innovation in sustainability and health. He is an associate of the International Institute for Sustainable Development and a research associate of the Centre for Urban Health Initiatives.

About the Contributors

Aiden Abram is a consultant with 2degreesC. Aiden has been consulting on climate change, water, and youth engagement work. Aiden brings skills and experience from six years of community leadership roles, and his formal education in earth science and international development.

Elisabeth Arnold is the Director of the FCM Centre for Sustainable Community Development. Prior to serving as an Ottawa city councillor from 1994 to 2003, she worked as a community developer. Elisabeth holds a Masters in Urban and Regional Planning from Queen's University.

Bill Beamish recently retired from the Town of Gibsons after 37 years in public service. He continues to work as a consultant to local government, and is interested in climate change mitigation and adaptation as it affects small coastal communities in British Columbia and elsewhere.

Lyle A. M. Benko is President of L*A*M*B* Environmental and Educational Consulting (Inc.) and currently Vice-Chair for the City of Regina's Green Ribbon Committee. He was elected as Co-Coordinator for the UN Regional Centre of Expertise Saskatchewan. Lyle serves on several boards at the local, provincial and national level that deal with environmental and sustainability issues.

Alex Boston was the David Suzuki Foundation's Senior Climate Policy Analyst and a British Council Scholar at the University of Oxford's Environmental Change Institute. He currently leads the Holland Barrs Planning Group's local government climate program.

Trevor Dixon Bennett is a consultant with 2degreesC. He joined the team in 2007, bringing a rich perspective from his intensive studies on climate change in Canada's North, complemented by his study and travel experiences in Iceland and Greenland.

Devin Causley holds a Masters of Applied Environmental Studies in Local Economic Development and a Bachelors of Environmental Studies in Planning from the University of Waterloo. He is a town planner by trade and a full member of the Canadian Institute of Planners. Devin currently works for the Federation of Canadian Municipalities as Senior Capacity Building Officer on climate change, energy and water issues. His contribution in Chapter 3 was developed on the basis of research and personal work experience.

Andrew Cowan is a Senior Manager at the FCM Centre for Sustainable Community Development. Previously, he was the Director of Climate Change for the Province of Manitoba and the Environmental Coordinator for the City of Winnipeg. Andrew holds a Masters in Natural Resources Management.

Clive Doucet has degrees in Urban Anthropology from the Universities of Montreal and Toronto. He's worked as an urban activist all his life, starting with the Stop Spadina Movement in Toronto when he was an undergraduate student in Toronto. He has written professionally in the field of urban affairs, as well as plays, novels, memoirs, and poems. His latest book is *Urban Meltdown: Cities, Climate Change and Politics as Usual*. He is married with two grown children and is now a grandfather of three.

Louis Drouin obtained his Masters in public health in 1978 and has specialized in community medicine since 1982. He is an associate professor in occupational and environmental health, and is currently the head of the urban environment and health sector of the Montreal public health department.

Sarah Eaton completed her BSc (Hons) and MSc in Earth Science at Memorial University of Newfoundland. Her interest in global climate change resulted from her Masters work with Canadian SOLAS (Surface Ocean Lower Atmosphere Study). She is currently working as a geologist with Crosshair Exploration and Mining in Labrador.

Michael Evans is an Edmonton-based strategic communications and public policy consultant with a particular interest in municipal environmental and sustainability issues. He has steered several public engagement strategies in the environmental realm.

Anne Golden, Ph.D., C.M., has been President and Chief Executive Officer of The Conference Board of Canada since October 2001. Previous to that, she served as President of The United Way of Greater Toronto for 14 years. Dr. Golden is the author of numerous publications on public policy issues.

Geraldine King is Manager, Environmental Initiatives, Engineering Department, City of St. John's, where she is responsible for promoting environmental awareness, and fostering activities that encourage environmental protection and enhancement in the city. Prior to her work with the city, she worked for the Province of New Brunswick in the Department of Environment.

Norman King obtained his MSc in epidemiology in 1978 and has been working in public health since then. He now works with the Montreal public health department, and his major fields of interest include indoor and outdoor air quality.

Tom Kosatsky, MD, is a Montreal-based community medicine specialist. He maintains a clinical practice in occupational and environmental medicine, and conducts research in environmental health for the Montreal public health department. During 2004, he was engaged as epidemiologist to the WHO European Centre for Environment and Health in Rome, with responsibilities in the area of climate change and health.

Caroline Larrivée is an infrastructure specialist with Ouranos, a research consortium on climate change. She studied Urban Planning at the University of Montreal.

Luc Lefevre obtained his Master's in toxicology in 1988 and since then his been working with the Montreal public health department. He is now the assistant coordinator of the emergency bureau and his major fields of interest are the emergency preparedness and response.

Dan Sandink, M.A., is Research Coordinator at the Institute for Catastrophic Loss Reduction. Through his research, he has authored numerous reports and articles on urban flood risk perceptions, and risk management and climate change adaptation.

Guillaume Simonet is an impacts and adaptation specialist with Ouranos. He recently began a PhD in Environmental Sciences and Sociology at Université du Québec à Montréal and Université Paris X on adaptation to climate change issues.

Edward Willett writes a weekly science column and is author of more than 30 books of both fiction and non-fiction, including several on science and engineering topics for both children and adults.

CLIMATE CHANGE RESOURCE SECTION

The world is changing.
Are you?

*Our clients
contribute to a
healthier and more sustainable world.*

*We help
by supporting their
leadership and innovation.*

www.2degreesC.com

Climate Change Central

Climate Change Central is a unique public-private partnership promoting development of innovative responses to global climate change and its impacts. Climate Change Central builds links and relationships between businesses, governments and other stakeholders in Alberta interested in pursuing greenhouse gas reduction initiatives.

Installation of a 1kW grid-connected solar PV system on the Jasper Activity Centre

The Alberta Solar Municipal Showcase project is an example of an exciting renewable energy demonstration involving 20 municipalities across the province showcasing grid-connected photovoltaic systems on highly-visible public buildings.

The Showcase is coordinated through Climate Change Central, spearheaded by the City of Medicine Hat and funded in part by the Federation of Canadian Municipalities.

For more information on this and other innovative initiatives, please visit **www.climatechangecentral.com**

MANAGING THE GHG CHALLENGE

CANADIAN STANDARDS ASSOCIATION

One resource. A world of solutions.

Canadian Standards Association can be your most important strategic resource for achieving your greenhouse gas (GHG) reduction goals.

CSA's Climate Change Services team provides industry and government with the products, services, training and strategic solutions needed to help you:

- Adapt infrastructure and built environment to meet the demands of a changing climate
- Understand how to measure, report and verify GHG emissions to meet ISO 14064 standards
- Showcase your GHG reduction activities
- Source valuable information about other carbon reduction projects on CSA's CleanProjects™ GHG registry
- Reduce costs through energy conservation and efficiency.

Find out more about how CSA can help your municipality improve its GHG footprint and reduce energy costs.

Visit www.csacanhelp.ca today.

® Registered trade-mark of Canadian Standards Association

DILLON
CONSULTING

*Serving municipalities across
Canada for over 60 years*

ADAPTATION TO CLIMATE CHANGE

INFRASTRUCTURE	ENVIRONMENT
RISK MANAGEMENT	STRATEGIC PLANNING
EMISSIONS INVENTORY	CARBON TRADING

Offices Across Canada and International
235 Yorkland Blvd., Suite 800, Toronto, Ontario M2J 4Y8
416.229.4646

www.dillon.ca

Building tomorrow, today.

www.sustainablecommunities.fcm.ca
Federation of Canadian Municipalities

FCM | Centre for Sustainable Community Development

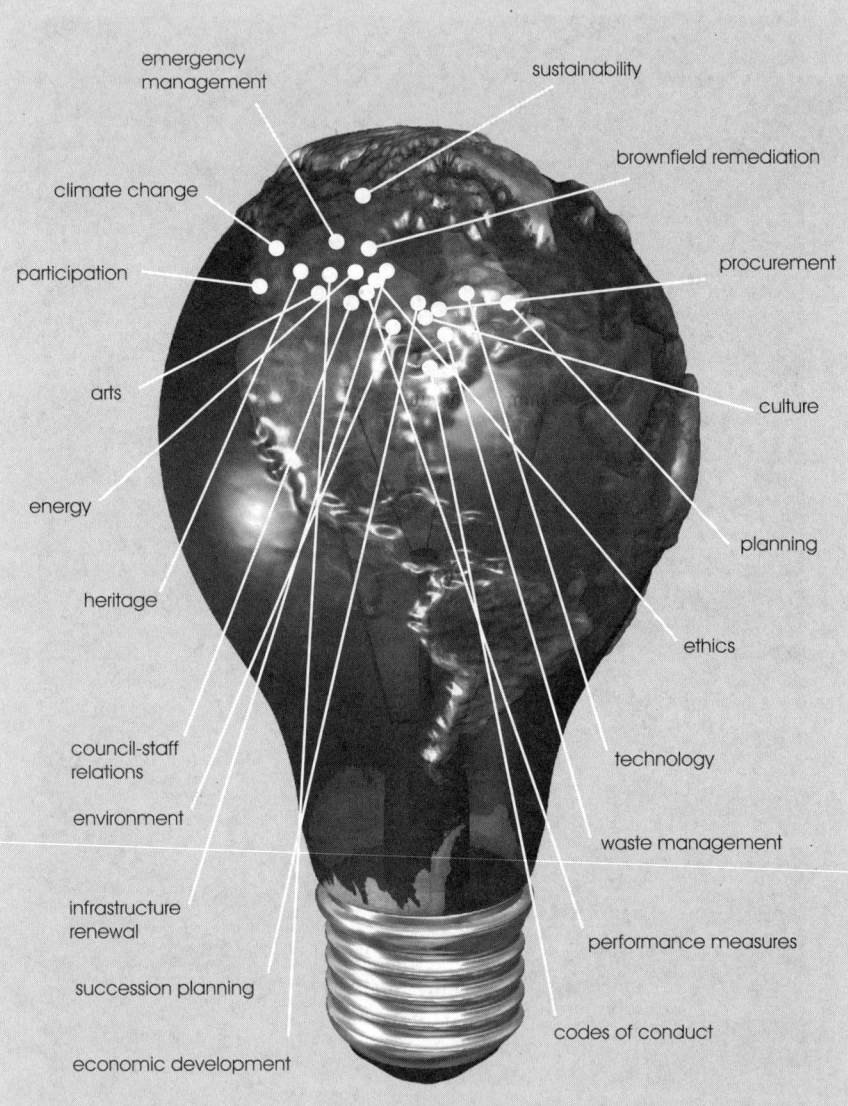

Building Sustainable Communities

At the Pembina Institute, we work on the issue of climate change from an integrated perspective. We have a dedicated team of planners, engineers, environmental scientists, policy analysts and political scientists that have experience developing climate change solutions for:

- Municipalities
- Corporations
- Individuals
- Provincial Governments
- The Federal Government
- Aboriginal Communities

This approach allows us to transfer learning and knowledge between the different groups, and build capacity to address climate change in an integrated manner.

Specific to municipalities, the Pembina Institute offers several custom services:

- Sustainability Planning
- GHG Planning
- Community Energy Planning
- Climate Change Adaptation Planning
- Smart Growth Planning
- Energy Policy Development

For more information on the work of the Pembina Institute, please go to: http://communities.pembina.org, or contact:

Jesse Row
Director, Sustainable Communities Group
Pembina Institute
(403) 269-3344 ext. 110
jesser@pembina.org

With offices in:
Vancouver, Calgary, Edmonton, Drayton Valley, Ottawa and Toronto

OTHER PUBLICATIONS FROM MUNICIPAL WORLD

To order any of the following Municipal World publications, contact us at: mwadmin@municipalworld.com, or telephone 519-633-0031 (toll free 1-888-368-6125).

By-law and Question Voting Law – Item 1288

Candidates and Electors – Item 1219

Deputy Returning Officers Handbook – 1280

Electing Better Politicians: A Citizen's Guide (Bens) – Item 0068

Guide to Good Municipal Governance (Tindal) – Item 0080

How to Campaign for Municipal Elected Office (Smither/Bolton) – Item 1284

Making a Difference: - Volume 1 - Cuff's Guide for Municipal Leaders (Cuff) – Item 0059-1

Making a Difference - Volume 2 - The Case for Effective Governance (Cuff) – Item 0059-2

Measuring Up: An Evaluation Toolkit for Local Governments (Bens) – Item 0061

Municipal Election Law – Item 1278

Ontario's Municipal Conflict of Interest Act: A Handbook (O'Connor/Rust-D'Eye) – Item 0050

Ontario's Municipal Act - codified consolidation – Item 0010

Open Local Government 2 (O'Connor) – Item 0030

Procurement: A Practical Guide for Canada's Elected Municipal Leaders (Chamberland) – Item 0070

Public Sector Performance Measurement: Successful Strategies and Tools (Bens) – Item 0060

Run & Win (Clarke) – Item 0020

Strategic Planning: A Users' Guide (Plant) – Item 0085

Truth Picks: An Observation on this Thing call Life (de Jager) – Item 0090